March of the Microbes

March of the Microbes

SIGHTING THE UNSEEN

John L. Ingraham

The Belknap Press of
Harvard University Press

Cambridge, Massachusetts
London, England

First Harvard University Press paperback edition, 2012

Library of Congress Cataloging-in-Publication Data

Ingraham, John L.
March of the microbes: sighting the unseen / John L. Ingraham.
p. cm.
Includes bibliographical references and index.
ISBN 978-0-674-03582-9 (cloth: alk. paper)
ISBN 978-0-674-06409-6 (pbk.)
1. Microbiology—Popular works. I. Title.
QR56.I54 2010
616.9'041—dc22   2009037712

For Nance

Contents

Foreword

Take a look at the world around you and what do you see? During a morning stroll in springtime we might, if not completely consumed by our inner conversations, take note of birds singing, traffic blaring, the peaceful solitude that accompanies sunrise in the forest. But at the contact of each foot with soil or concrete and as we involuntarily take in each breath of air, we seldom consider the invisible yet vast world of microbes that we walk on, that surround us, and that we even harbor in our bodies. Much less do we consider the impact that these tiny creatures have had on Earth, a planet largely dominated by microbes, where humans are but a small minority of its inhabitants. These invisible creatures have occupied our planet for at least 3 billion years and accompany us from birth until death. Our constant companions, they have been present throughout our evolutionary history.

No matter in what surroundings we find ourselves, we will be witnessing the effects of microbial activity. Imagine sitting at the beachside watching the stunning spectacle of a sunset over the ocean, with clouds brightly illuminated by the disappearing sunlight. Now let your thoughts go to the air you breathe and consider that microbial activity produced a significant proportion of the life-sustaining oxygen you take in with every breath. Go beyond the oxygen and consider the inert nitrogen gas that makes up the majority of the air we breathe. That nitrogen needs to be converted into a form that all plants and animals can assimilate. This ability to turn inert nitrogen into an available nutrient is no easy task, yet it was mastered long ago by microbes. In fact, all the major chemical cycles in our planet have a microbial component. Even the clouds so beautifully illuminated at sunset were produced as a result of the action of microbes in the oceans' waters. Perhaps this should not be surprising. Every tiny corner of Earth's biosphere is filled with microbes. Within our view in that idyllic beachside setting there are sand and water aplenty. Every grain of sand and every drop of water is teeming with microbial life—life so rich and so mysterious that we are just beginning to comprehend the extent of its diversity. There is still so much to explore in the microbial world!

Thus, if we are to begin to understand our surroundings better we need to gain knowledge of how these microbes' ubiquitous activities can so profoundly influence our existence. Where to start? Read on. In your hands you hold a field guide to the world's microbes. This magnificent primer has been given to us by the one person I would have chosen to take on the task. John Ingraham has been, for more decades than he or I would care to acknowledge, a passionate observer of the microbial world. John has not only explored the activities of microbes as a first-rate academic researcher, he has written several key textbooks widely used in the teaching of microbiology around the world. In composing this volume he has drawn on his extensive experience as a naturalist and a writer. The result is a narrative that will open your mind to the wonders of the invisible world of microbes.

I was fortunate enough to work in the same laboratory as John when he was on sabbatical at the University of California at San Diego and I was a graduate student. In his characteristically unassuming yet powerful way he shared his extensive knowledge and awoke in me the desire to explore the splendor of the microbial world. I have been fortunate to participate in this exploration ever since. But I have been even more fortunate to have John be both an inspiration and a lifelong friend. As I read through the journey that you are about to embark on, and which will take you through dozens of microbial sightings, I was repeatedly awestruck and felt goose bumps in a way that only a master teacher can bring upon the student. I am sure you will thoroughly enjoy the expedition.

Roberto Kolter
President, American Society for Microbiology

1

The Microbial Landscape

Microorganisms—microbes—are our progenitors, our inventors, and our keepers. A few of them occasionally become our adversaries, even our killers. They might even make us fat. But we are the recent intruders into their well-established and self-developed world. They appeared on Earth about three and a half billion years ago, just one brief thousand millennia after the planet formed. We—*Homo sapiens*—appeared only a little over 100,000 years ago and our near ancestors a few million years before that. We have journeyed with microbes a few steps on their solitary travels: not quite two inches of their mile, or two and a half seconds of their day. Plants and animals in their most primitive forms have inhabited Earth only a little over a quarter as long as the microbes.

Such established settlers deserve a certain measure of respect. But microbes did much more than arrive first. They fundamentally changed the chemistry of the planet in many ways, rendering it hospitable enough for us to evolve and exist here. They produced the atmospheric oxygen we breathe as well as the chemical forms of nitrogen essential for the plants and animals we eat. And microbes continue to maintain our environment in life-sustaining balance in spite of our massive assaults on it.

Because their form and size are so unimpressive, we are likely to conclude that microbes' evolutionary accomplishments are trivial. Certainly, this is not the case. They solved all the fundamental problems of life—genetic, metabolic, and structural. We just added the more obvious, and perhaps dangerous, embellishments such as consciousness, intelligence, and articulate speech. They learned to write and read the universal genetic code, using it to make the macromolecules—proteins, nucleic acids, polysaccharides, and lipids. Microbes learned to assemble these macromolecules into membrane-enclosed structures called *cells,* the fundamental units from which all life is built. Macromolecules comprise and run all cells, including our own. As modern genetic techniques have unequivocally demonstrated, we acquired the basic skills of life by inheriting genes from microbes. We stand at the end of the long evolutionary trail they began to blaze.

But microbes did not share all their talents. As though to ensure our continued dependence on them, they kept some of their metabolic skills exclusively to themselves. These include some of their

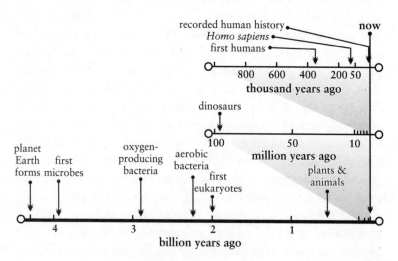

Histories of Earth and life.

unique abilities to keep Earth habitable. Only microbes, for example, can fix (utilize the gaseous form of) nitrogen, return gaseous nitrogen to replenish the atmosphere, and degrade cellulose, the major structural component of plant cells and the most abundant organic nutrient on Earth. All plants and animals, including ourselves, are directly or indirectly dependent on these microbial transformations. We would soon disappear without the continued intervention of microbes.

Of course, some microbes harm us: damage our crops, harm our domesticated animals, spoil our food, cause disease, and sometimes kill us. But harmful microbes are a minuscule minority. Indeed, the percentage of disease-causing microorganisms (pathogens) is far, far less than the percentage of humans that commit first-degree murder. Yet, even though microbial felons are few, their impact on our evolution, history, and survival has been profound. Of necessity, animals evolved an elaborate set of defenses, including the immune system, to protect themselves against microorganisms. And that is no small consideration. Much of our body and many of our genes are dedicated to running the immune system.

If our immune system becomes unable to function for any reason—hereditary, environmental, or pathological—we soon become overwhelmed by microbes, and we quickly die. Recall the "boy in the bubble," born with a massively deficient immune system. He was able to survive only in the artificial, sterile environment of his bubble, completely isolated from the microbe-filled world. More commonly, untreated AIDS, which damages one vital component of our immune system, renders us vulnerable to other microbes that administer the coup de grâce.

In spite of the importance and intrinsic fascination of microbes, many of us with an otherwise abiding interest in the natural world largely ignore them, probably for one overriding reason: we cannot see most of them. We cannot be microbe watchers with the ease that we become bird watchers. We cannot identify a microbe by sight as we would a rare orchid. A high-quality microscope is es-

sential for watching individual microbes such as bacteria. Even
then, the rewards are minimal. Most bacteria look alike except for
differences in shape, which are readily apparent microscopically.
But even shape does not help much for identification. A highly experienced microbiologist would still have great difficulty distinguishing between similarly shaped microbes under the microscope.
Vibrio cholerae, which causes epidemics of deadly cholera, and
Alivibrio fischeri, which does nothing more threatening than light
up parts of certain marine animals, appear to be identical—both
are rod-shaped cells with a flagellum at one end.

Their activities, not their appearance, are what distinguish one
microbe from another. Many of the consequences of their activities
are readily apparent to our eyes, nose, taste buds, and sometimes
our ears. Our unassisted senses can tell us much about what microbes are and what they do. Throughout this book, we will explore the easily discernible consequences of microbial activities,
which will reveal something about the microbes themselves. Why
shouldn't microbe watching be as accessible and enriching as bird
watching? We might even want to start a list of microbe sightings,
in the manner of bird watchers. A walk in the woods becomes much
more interesting if, in addition to the animals and plants we can
identify, we notice the vital role microbes play in maintaining these
other forms of life. Some microbes are invisible to our senses, yet
they have major impacts on our environment and our lives. We will
discuss some of these wallflower microbes as well.

Bird watchers stretch the usual meaning of "sighting" to include
hearing: recognizing a birdcall counts. As microbe watchers, we
will stretch the meaning further to include smelling, feeling, and
even tasting as well as seeing and hearing. We will use all our
senses.

After a bit of microbe watching, we become more attuned to the
impact of these small creatures, neglected and disrespected by even
seasoned, knowledgeable naturalists. For example, the otherwise
admirable timeline on the great sweeping wall in front of the visi-

tors' center at Dinosaur National Monument on the border between Utah and Colorado begins at Earth's formation and ends with the present, chronicling the appearance of various life forms along the way. Shockingly, it completely lacks any reference to microbes. Microbe watchers know that the life forms shown on the wall could neither have evolved nor survived without microbes. Neither could the ones examining the wall.

Experienced microbe watchers will be disappointed with signs such as the one in the Desert Botanical Garden in Phoenix, Arizona, a beautiful and richly stocked treasure, that blatantly proclaims, "Only plants make their own food." Even novice microbe watchers would counter that microbes were the first to make their own food from atmospheric carbon dioxide and many continue to do so—some by using light and some by using the chemical energy locked in mineral materials such as forms of iron, sulfur, and

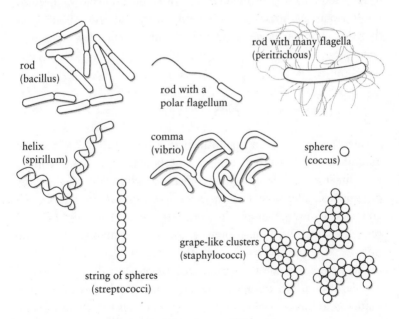

rod
(bacillus)

rod with a
polar flagellum

rod with many flagella
(peritrichous)

helix
(spirillum)

comma
(vibrio)

sphere
(coccus)

string of spheres
(streptococci)

grape-like clusters
(staphylococci)

Some common shapes of Bacteria.

manganese. Plants themselves depend on microbe-derived genes to make their food. Indeed, they employ long-ago captured photosynthetic microbes, now apparent as intracellular structures called *chloroplasts,* to do so. The sign more accurately should announce, "Only microbes make their own food. Plants obtain their food by exploiting microbes as intracellular, food-producing slaves."

At Lake Abert in southeastern Oregon, an informative nature sign describes and illustrates the birds you might see and the underwater brine shrimps you cannot but does not mention the black, extensive community of microbes in full view right behind the sign. Many visitors must wonder what the big black area is. Microbe watchers would certainly know. Microbe sightings can be diverse: at a picnic at Lake Abert, we might notice the black spot while wondering why the Swiss cheese in our sandwich has holes. Every day we notice many similar phenomena, but we rarely relate them to microbes. To begin to understand these phenomena, we need to know what microbes are, what kinds exist, and a bit about what they do. That means we ought to be familiar with a few terms and principles.

Some Definitions

A microbe is simply an organism that is too small to see without magnification. Most of us can see objects that are about a tenth of a millimeter (or a hundred micrometers, 100 μm) in size, so any organism smaller than 100 μm is a microorganism, a microbe. Of course, organisms that are the same size are not necessarily closely related. A frog and a mushroom are about the same size. We should not be surprised, then, to learn that microbes are all small but highly diverse. In fact the various groups of organisms that qualify as microbes are as distantly related to one another as they are to plants or animals. We know this because modern methods of comparing the forms of macromolecules that occur in organisms have

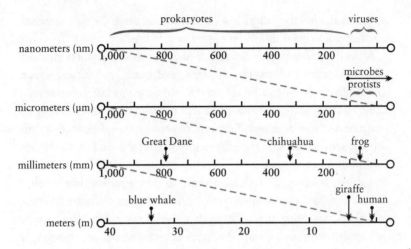

Relative sizes from viruses to blue whales.

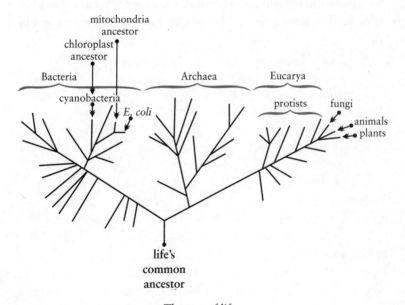

The tree of life.

revealed the relationships among all living things, often summarized as a universal family tree—the tree of life.

All cellular organisms belong to one or another of three distinct groups (domains): Bacteria, Archaea, and Eucarya. Viruses, which are acellular, do not show up on the tree of life at all; nevertheless, some of them appear in this book. All plants and animals belong to just one of these domains, Eucarya, the eukaryotes. In contrast, microbes are spread among all three. All Bacteria and Archaea are microbes, in fact.

Because bacteria and archaea share the same rather simple cellular architecture (in contrast to the more complex cellular structure of eukaryotes), they are collectively called *prokaryotes*. Prokaryote is a useful term, but it only describes appearance; it does not imply relationship. Representatives of the two prokaryotic groups, bacteria and archaea, are indistinguishable under the microscope. But in spite of their looking alike, the microbes belonging to these two groups are utterly different genetically and biochemically. They are as distantly related to each other as either group is to plants or ani-

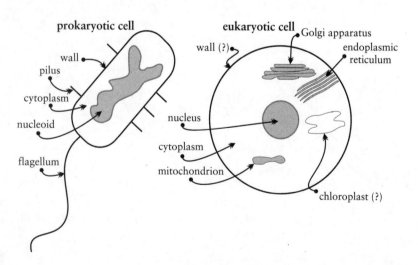

Life's two kinds of cells.

mals. It is not improbable that two groups of organisms sharing the same (prokaryotic) cell structure are so different because prokaryotic cells are characterized by their simplicity: what they do not have, not what uniquely evolved structures they came to share.

Bacteria

Bacteria is a familiar term. It describes that large group of mostly single-celled prokaryotes with well-known names such as *Escherichia coli, Staphylococcus aureus,* and *Streptococcus pneumoniae* that we might read about on the back of a can of disinfectant spray. Bacteria, known by many as "germs," include some representatives that cause diseases, including tuberculosis, pneumonia, strep throat, and, as we have somewhat recently learned, peptic ulcers.

Bacteria are divided into two major groups—gram-positive and gram-negative. These names derive from a staining procedure developed and named over a century ago by a Danish microbiologist, Hans Christian Gram. The stain is selective in that only certain bacteria are colored by it. The bacteria that are colored came to be designated as gram-positive. The other bacteria that the stain does not color were designated gram-negative.

Staining (adding dyes to bacterial cells to be viewed under the microscope) was taken very seriously by microbiologists in Gram's time, because unstained bacterial cells were very difficult to see using the microscopes then available. Individual bacterial cells do not absorb much light. Because light passes right through them with almost undiminished intensity, they are nearly invisible when viewed with conventional microscopes. But stains are effective absorbers of light, so those that attach to or enter bacterial cells render them crisply visible. Certain stains act selectively on different bacteria, as Gram's stain does. Advances in the technology of staining propelled the spectacular scientific accomplishments of the late nineteenth century, known as the "Golden Age of Bacteriology"—a brief period during which the particular microbes that caused most known bacterial diseases were identified.

As more and more bacteriologists dedicated themselves to staining, techniques became more sophisticated and useful beyond just rendering microbes visible. Highly selective stains were discovered. Some, for example, could color a particular bacterium but not the human cells it was infecting. Development of this sort of stain enabled Robert Koch, a German bacteriologist, to discover in 1882 the causative agent *(Mycobacterium tuberculosis)* of tuberculosis. Using his new acid-fast stain, he was able to correlate having the disease with the presence of this deadly bacterium. Other stains that selectively color individual internal parts of bacteria, revealing their presence or absence, became valuable tools for learning how bacterial cells function.

Christian Gram's stain accomplished something even more remarkable. As stated, it divided bacteria into two major groups: gram-positives and gram-negatives. Soon microbiologists learned that these two groups differed in many respects. Knowing whether a bacterium was gram-positive or gram-negative became valuable information. But it was not until the mid-twentieth century, when electron microscopes became available, that the structural differences between gram-positive and gram-negative bacteria were laid bare.

The distinction is anatomical: all cells are enclosed within a single, highly organized lipid (fatty) layer called the cytoplasmic or cell membrane, but gram-negative bacteria have an additional external membrane, the outer membrane. This curious structure, which is chemically similar but distinct from cell membranes, provides gram-negative bacteria with a unique protective shield against toxic agents in the environment, rendering these bacteria thousands of times more resistant to certain antibiotics, for example. The double membrane structure confers another distinctive capability to gram-negative bacteria. Its presence creates a unique intermembrane region called the *periplasm,* which acts as a digestive chamber. Certain larger molecules that cross the outer membrane to enter the periplasm are digested there into smaller component molecules that can cross the cell membrane and nourish the cell.

The gram-positive/gram-negative distinction correlates with a host of other properties as well: relatedness and sensitivity to antibiotics being just two of them. You have learned a lot when you are told whether a bacterium is gram-positive or gram-negative. For example, it is one of the first things a physician wants to know when treating a bacterial infection. But in spite of the striking correlation between Gram staining and presence of an outer membrane, that is not the basis for the differing reaction to the stain. Instead, it is another difference in the cell's covering layer, the cell wall. The thicker, chemically more complex wall of the gram-positive bacteria cell retains the stain; the thinner gram-negative wall does not.

One group of gram-negative bacteria, cyanobacteria, deserves special attention. These bacteria are ubiquitous in sunlit environments and are major actors in many ecosystems because they are so versatile: not only are they capable of photosynthesis, most of them are able to utilize nitrogen gas from the atmosphere as a nutrient. So all they need to grow and prosper is sunlight, air, and a few minerals. Cyanobacteria also played a key role in Earth's evolutionary development. One of them was the bacterium captured by a progenitor plant cell, which conferred on it and its progeny the capacity for photosynthesis. This event made the plant world possible. And the type of photosynthesis that cyanobacteria and plants carry

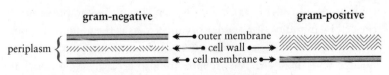

Envelopes that cover bacteria.

out produced, as a byproduct, all the oxygen in Earth's atmosphere. This momentous contribution made life possible for us and other animals.

Because the cells of cyanobacteria are about ten times larger than those of most other bacteria, microbiologists once thought they must be something else, and called them "blue-green algae." But their cell structure is clearly prokaryotic; algae are eukaryotic. So cyanobacteria are just large bacteria, now sometimes called blue-green bacteria, but ecologically important ones.

Archaea

What about the archaea, the other and very different group of pro-karyotes that constitute a domain of their own? Various species of this group have been known since the nineteenth century, but they were not recognized as being distinct until relatively recently. They were considered to be just another group of bacteria—nothing more, nothing less—because they look like bacteria and have the same prokaryotic cell structure. In the 1970s, however, with the advent of new, powerful tools such as DNA sequencing to study relatedness of living things, an American microbiologist, Carl Woese, discovered that this group of prokaryotes is only distantly related to bacteria and, indeed, to all other living things. Subsequent studies revealed startling biochemical distinctions and other characteristics that make them unique.

For example, the chemical composition of their cell membrane (that enclosing structure mentioned earlier that surrounds the protoplasm of all cells) is radically different from the composition of the cell membranes of all other organisms, as is the composition of certain of their key enzymes, such as the one that makes their ribonucleic acid, or RNA. The archaea are also ecologically and metabolically unique. Only they have representatives (the methanogens) that make methane (natural gas). Cows belch methane because they have these archaea in their rumen, and the intestinal gas of some humans is methane for the same reason: harboring archaea. Various other groups of archaea are capable of inhabiting and thriving in what we would consider extremely hostile, even lethal, environments. Archaea hold the biological records for ecological toughness and adventurism. Some thrive at temperatures well above the boiling point of water, up to at least 113°C; others make their homes in saturated brine solutions; still others are found in environments as acidic as stomach acid (pH 1.0). But not all archaea are restricted to such extreme environments. Some live in Earth's most temperate and hospitable environments. The oceans, for example, are home for large numbers of archaea.

Unlike members of other groups of microbes, no archaea are known to cause diseases of humans, animals, or plants. Some microbiologists speculate that archaea are so ancient that they evolved and established their ecological preferences long before animals or plants made an appearance. Archaea found other habitats and stayed there. The theory meshes with finding many archaea in environments that are characteristic of the early conditions of our planet—too hostile for most other microbes, which evolved on a more benign Earth.

Eukaryotic Microbes

All microbes other than bacteria and archaea are eukaryotes. As I stated earlier, unlike prokaryotes, not all eukaryotes are microbes. A eukaryotic cell possesses what a prokaryotic cell lacks—a nucleus.

Fungi. One group of eukaryotic microbes is fungi, which include such organisms as mushrooms, molds, and yeasts. These constitute a tight evolutionary cluster of organisms characterized by a distinctive cell structure—long cytoplasm-filled tubes not divided by complete cross walls, although some members of the group, the yeasts, evolved to exist under certain conditions as single cells. Fungi play critical environmental roles. They are the principal decomposers of plant debris. Leaves and branches that fall in the forest are recycled for the main part by fungi. Some fungi also attack living plants. Indeed, fungi are the major agents of plant disease. But they are not the dominant agents of animal disease; bacteria and viruses are. Some fungi do infect animals, including humans. Athlete's foot and ringworms are annoying but not life threatening; others such as San Joaquin Valley fever and blastomycosis most definitely are. You might be surprised to learn that these diseases are notoriously difficult to treat because fungi are so closely related to us and metabolically similar. It is difficult to kill them with drugs without doing great damage to ourselves.

Fungi are divided on the basis of their sexual spores into three groups—Ascomycota, Basidiomycota, and Fungi Imperfecti. The ascomycetes package their spores in a sac called an *ascus;* the basidiomycetes arrange theirs on a clublike structure called a *basidium.* Nineteenth-century mycologists also identified a group of fungi that seemed to lack a sexual stage or did not make sexual spores, which they termed "imperfect." There are at least two reasons why an organism might lack a sexual stage: it might be genetically incapable of having one, or it might not have found a suitable partner. In either case, there is no way to know whether a fungus is an ascomycete or basidiomycete if it lacks a sexual stage and, therefore, cannot form sexual spores. So such a fungus reproduces only asexually and is consigned to a provisional group, the imperfect fungi, until its sexual stage is observed, if it is. Then it is given a new name.

Protists. Nonfungal unicellular eukaryotic microbes were traditionally divided into two groups, protozoa and algae. The photosynthetic ones were designated as algae and the nonphotosynthetic, the largely motile ones, as protozoa. The motile, photosynthetic organisms such as *Euglena* spp. occupied an awkward middle ground. But recent molecular studies on relatedness destroyed this traditional classification scheme. The unicellular algae are not closely related to multicellular algae, and there is no clear distinction between them and the organisms called protozoa. Both unicellular algae and protozoa, along with some organisms formerly considered to be fungi, are now included in a group called the *protista* (protists), an old microbiological term that has been pressed back into service. Some groups of protists, such as the dinoflagellates and diatoms, change the environment. Others cause serious human diseases such as malaria. Because protist-caused diseases are particularly prevalent in tropical climates, their study traditionally was called tropical medicine. But protists such as *Giardia* and *Cryptosporidium* also cause diseases in temperate regions.

Viruses

Members of all three domains of living things—Bacteria, Archaea, and Eukaryotes—share the characteristic of being cellular organisms. All are composed of one, a few, or a huge number of packages of protoplasm (cells), each enclosed within an intact cell membrane. Viruses are completely different. They are not made up of cells. Instead, viruses are mere packages of genetic information, stored in the sequences of the component parts of the two kinds of nucleic acid also found in cells, either DNA (deoxyribonucleic acid) or RNA (ribonucleic acid), never both, and packaged in a covering of protein. When either kind of viral nucleic acid enters a cell (that is, when a virus infects a host cell), its encoded information directs the cell to start making more packages of viral components, sometimes resulting in damage to or the death of the host cell and sometimes just persisting there without causing much difficulty for its host. Viruses are the ultimate parasites. They do nothing themselves; they just command cells to do their bidding.

One type of virus, bacteriophage T4, provides a good example of what viruses do. Bacteriophage T4 happens to kill its bacterial victim.

The phage particle (virion) attaches by its tail to the surface of a bacterial cell and "injects" the DNA within it into that cell. Only the phage's DNA (information content) enters the cell. The rest of the phage (composed of protein) stays outside as an empty hulk attached to the cell surface. That simple event tells the fundamental story of viruses. Unlike cellular organisms, which remain intact as they reproduce, viruses disassemble when they infect a host cell to begin their cycle of reproduction. The virion itself is inert. It has no metabolism of its own and cannot reproduce without infecting a cell.

When the DNA from a phage T4 virion enters the host cell, it orders the cell to stop what it is doing and direct all its metabolic activities instead to making components—DNA and protein—of more virions. Then, when the host cell has made enough of these

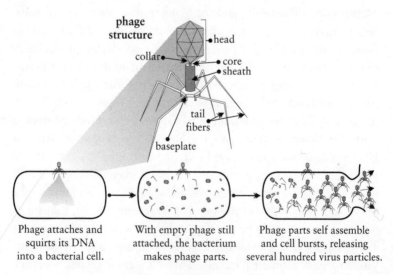

Phage attaches and squirts its DNA into a bacterial cell.	With empty phage still attached, the bacterium makes phage parts.	Phage parts self assemble and cell bursts, releasing several hundred virus particles.

Life of bacteriophage T4.

phage components, they spontaneously assemble into new phage virions. About twenty minutes after the cell was infected, it disintegrates (lyses), releasing some 200 phage virions, each capable of infecting another bacterial cell. The growth cycle of phage T4 (and many other viruses as well) is almost explosive. Twenty minutes after the initial infection, one phage virion becomes 200. If each of them infects yet another bacterial cell, after forty minutes the original phage particle will have 40,000 progeny, and after an hour, 8 million.

There are many variations on the general theme of viral growth. Some viruses contain RNA instead of DNA. Many do not kill the host cell they infect. Some do not start to reproduce immediately after infection. But the mode of reproduction of phage T4 does illustrate the fundamental principles of viral reproduction: the virion disassembles as it infects, and viral nucleic acid (DNA or RNA) then directs the infected cell to act in the virus's, not the cell's, best interests.

Some microbiologists and probably microbe watchers as well indulge passionately in tiresome arguments about whether or not viruses are alive and therefore have the right to be included in good standing among the microbes. Viruses are studied in university departments of microbiology and they do manifest many of the characteristics of living things: they reproduce, they evolve, they use the genetic code, and they are composed of the same macromolecules found in all living things. In contrast, cellular organisms store their genetic material in DNA, which directs their activities, and, while alive, never disassemble.

Growth and Reproduction of Cellular Microbes

The principal activities of microbes other than viruses are easy to summarize: chemical syntheses of more microbial cells. We often think of this process more familiarly as growth and reproduction.

Of course, growing and reproducing are prime, essential characteristics of all life. But in higher forms of life, as we like to think of
ourselves, they do not dominate. Most of our metabolic energy is
spent in other ways, such as movement, thought (our brains use
about 20 percent of our total available energy), and maintaining
body temperature. In contrast, if conditions are favorable, almost
all of a bacterium's metabolic energy is spent making more bacterial cells. Even when actively swimming, which some bacteria can
do very well, only a marginal percentage of their metabolic energy
budget is diverted from growth and reproduction into motion.

As a microbial cell makes more cells like itself, it changes its environment in a variety of ways: by using up nutrients, by forming a
greater mass of cells, and even more so by releasing metabolic waste
products into the environment. We notice the activities of microbes
by the friendly fire they spray while pursuing their dedicated objec-
tives—growth and reproduction.

Microbial growth and reproduction are chemical processes. Raw
materials from the environment are taken into cells and there transformed into the stuff needed to make more cells. This transformation depends on a complex set of chemical reactions—in most
microbial cells, around a thousand different ones. Each of these reactions (with a very few minor exceptions) is facilitated by a different biocatalyst or enzyme. So thousands of different enzymes are
needed to make a microbial cell.

In spite of its complexity, the overall process of reproduction can
occur with astounding rapidity. A cell of an ordinary bacterium
such as *Escherichia coli* (possibly the most meticulously studied
and thoroughly understood of all living things) can, under ideal living conditions, accomplish this labyrinthine chemical feat of becoming two cells in about 15 minutes. A few bacteria can do it just
a little more quickly, although many microbes require much longer.
The few thousand individual chemical reaction steps (*E. coli* has
1,276) involved would seem to be tremendously difficult to understand. However, when viewed not individually but as sets of pro-

cesses, rather like looking at the steps in an assembly line instead of the individual movements of the people who work there, the process becomes startlingly simple. Only five such unit operations or metabolic tasks are needed to make a new bacterial cell (or for that matter, any cell).

Acquiring nutrients. The first of these unit operations might seem familiar to us. Any of a great variety of raw materials (or what we call food) that might be available in the environment are taken into the cell. Usually this taking up is an active, energy-requiring process that concentrates nutrients, sometimes thousandsfold, inside the cell, thereby allowing microbes to thrive in environments in which the concentrations of nutrients are extremely low. Representatives of one genus of bacteria, *Caulobacter* for example, can thrive in tap water, which is not exactly a nutrient-rich soup. Yet there are enough nutrients there for the particularly miserly *Caulobacter* to scavenge, concentrate, and grow on.

Processing nutrients. After the ingredients are acquired, nutrients are processed into a common set of chemical starting substances inside the cell. Remarkably, in most cases only twelve different organic chemicals, called *precursor metabolites* (PMs), are needed to make any and all cells. Some prokaryotes require a thirteenth to make their unique cell-wall substance—peptidoglycan.

Making the building blocks of macromolecules. Next, some of the PMs are converted by a network of chemical processes into the approximately fifty different small molecules or building blocks, groups of which become linked together to make the cell's complement of macromolecules. All organisms require essentially the same set of building blocks. For example, all organisms require the same twenty amino acids, which are strung together in various sequences to make their proteins. But not all organisms are able to manufacture these building blocks on their own. Their ability to make their own building blocks depends on which enzymes they possess. *Escherichia coli,* as well as many other microbes, can make all twenty amino acids because it has all the necessary enzymes to do so. In contrast, we humans lack many of these enzymes. We can only make ten amino acids and, therefore, must have the other ten of them—the essential amino acids—made available in our diets.

Making macromolecules. The building blocks, usually called *monomers,* are strung together, or polymerized, into long chains to form the cell's macromolecules. The macromolecules of a cell are principally proteins, nucleic acids (DNA and RNA), polysaccharides, and lipids. Each will play a role in building the organism.

Assembling macromolecules. Macromolecules are then assembled to form the new cell's visible parts (organelles). When enough macromolecules have been formed and enough new parts have been assembled to double the cell's complement of them, the cell divides and the process is complete: a single cell has processed enough raw material to become two cells.

Unsurprisingly, the five metabolic tasks are not completely sequential or linear. Several feedback loops link them. The most important of these loops involves energy. Nature, as most of us have come to realize, does not provide free lunches. Chemical reactions, including those that occur in cells, proceed only if they pay their way with chemical energy. Some of the chemical reactions that occur in a cell are energetically self sufficient, but other essential ones are not. They need outside help.

How does the cell provide this help? Quite simply. The cell saves the excess energy from some of the metabolic reactions that occur on their own and uses it to push the others that need help. The cell stores this excess energy in the form of a remarkable compound called *adenosine triphosphate* (ATP). The excess energy of reactions that proceed on their own is trapped and used to form ATP, and later that ATP is consumed by other reactions that need assistance. ATP is the currency of metabolic energy, passed back and forth between energetically affluent and impoverished reactions.

So from a process point of view, the synthesis of a new cell is not that complicated: a cell uses nutrients from its environment to make the twelve precursor metabolites (PMs) and accumulate metabolic energy (ATP). Then, the cell uses these two vital metabolic resources to make a new cell.

All organisms, from microbes to humans, convert these metabolic resources (PMs and metabolic energy) into new cells by essentially the same chemical routes and the same sets of enzymes, but considerable variation exists among organisms—particularly among microbes—in the way they make their starting materials. The diversity of ways that microbes make PMs and accumulate metabolic energy is resplendently vast, and observing this diversity is fundamental to distinguishing one microbe from another. We can see, smell, taste, and sometimes feel or hear the consequences of the ways that microbes make PMs and accumulate metabolic energy.

In contrast, humans and other animals are restricted and prosaic accumulators of metabolic resources. We make PMs from the organic ingredients of the food we eat and acquire metabolic energy by aerobic respiration, a process similar to burning. The carbon atoms of organic nutrients are oxidized to carbon dioxide, and oxygen from the air we inhale is reduced to water, thereby releasing metabolic energy.

Plants are a bit more inventive. They make their PMs from atmospheric carbon dioxide and acquire metabolic energy from light by oxygenic (oxygen-yielding) photosynthesis. In other words, light energy and water yield oxygen gas and metabolic energy. At night they resort to aerobic respiration at the expense of some of the organic nutrients that they acquired during the day.

As we will see, microbes are almost unbelievably innovative and diverse in the ways they acquire their metabolic resources. Various microbes make their PMs from carbon dioxide, carbon monoxide, or organic nutrients, but a much more extensive collection of them than animals can use. For metabolic energy, they can respire organic compounds in several ways—not just aerobically; they can also apply this armamentarium to respiring inorganic compounds; they can utilize several forms of photosynthesis—not just oxygenic; they can ferment; and some can acquire energy just by shuffling certain compounds in and out of their cells. It is said, although it is a bit hard to prove, that no natural (not human-made) organic compound exists that cannot be used by some microbe as a source of its PMs, and no natural chemical reaction exists that cannot be exploited as a source of metabolic energy. No wonder microbes are so ubiquitous, that they occur in such huge numbers (there are 10 to 100 trillion individual microbes in and on you right now), and that they have such profound impact on us and our world.

Now we will start some actual microbe watching, smelling, tasting, or hearing. We do not need a microscope or other instruments for this enterprise. Our five unaided senses will do. We will begin with a sighting that depends on our sense of smell, which is startlingly sensitive, being able in some cases to detect just a few molecules. Smell can tell us a lot about what microbes are doing.

2

Acquiring Metabolic Energy

As I have just noted, our ability to perceive many microbes is a consequence of the ways that microbes change their environment as they set about acquiring metabolic energy. The four in this chapter are no exception; in fact, these microbes acquire energy in entirely unique ways.

A Smelly Rock Cod

What's that smell? We can't see a less-than-completely-fresh ocean fish begin to change, but its distinctive fishy odor assures us that there are microbes at work, and it tells us what kind they are as well as what they are doing.

When fish are first taken from the ocean, they are virtually odorless. But unless they are refrigerated immediately, they rapidly begin to smell fishy—perhaps pleasantly—within an hour or so. Soon, however, the odor turns offensive, even overpowering. Not so with freshwater fish. They smell okay until they rot. The fishy odor that develops in ocean fish comes from an intensely odoriferous substance, trimethylamine (TMA), that some bacteria make as they grow on a fish. Because bacteria that can make TMA are every-where (even the well-known *Escherichia coli* that lives in our intes-tines can make it) and because fish are a rich source of nutrients, some bacteria that can make TMA will surely be present. They be-gin to proliferate as soon as the fish dies. Because TMA has such an intense fishy odor, we notice the smell long before any other signs of decay appear.

Why do these bacteria make this odor and why only when grow-ing on ocean fish? They do so in their process of acquiring meta-bolic energy. As soon as an ocean fish dies, bacteria begin to grow on its surface as they would on any dead thing. Initially, they ac-quire their metabolic energy by aerobic respiration. By doing so, they quickly deplete the supply of oxygen in the nooks and crannies on the fish's surface. For us and most all other animals, that would be the end. Without oxygen we could no longer generate metabolic energy, and our metabolism would stop. Indeed, we and most ani-mals would quickly die.

When microbes run out of oxygen, however, many can switch, almost instantly, to some oxygen-independent means of acquiring metabolic energy. The bacteria growing on fish do this by substi-tuting a compound called trimethylamine N-oxide (TMAO) that is abundant in ocean fish for the missing oxygen. This resulting oxygen-independent form of respiration is called *anaerobic respira-tion*. Organic nutrients and TMAO are metabolized to form carbon dioxide (CO_2) and TMA, and release metabolic energy.

As with aerobic respiration, the organic nutrients in the fish are oxidized to carbon dioxide and metabolic energy is generated, but rather than forming water by reducing oxygen, smelly TMA is formed by reducing TMAO. Anaerobic respiration is by no means limited to substituting TMAO for oxygen. Many other oxygen sub-stitutes are utilized by various microbes. Indeed, some microbes can use several different substitutes. They choose oxygen or one of its substitutes strictly on the basis of energy economics and, of course, availability. Oxygen is a more powerful oxidizing agent than TMAO, so aerobic respiration yields more metabolic energy than TMAO-dependent anaerobic respiration. Microbes that are capable of both forms of respiration can sense this difference, and they utilize oxygen until it is depleted. Only then do they switch to TMAO-dependent anaerobic respiration. Microbes that can use several oxygen substitutes use the most energy-rich form of respiration available to them, switching from one to the next as richer substitutes become depleted.

Why do only ocean fish smell fishy? That is because only they contain the trimethylamine N-oxide that is reduced to smelly TMA. Freshwater fish lack it. TMAO that accumulates within the cells of ocean fish benefits the fish in several critical ways. It helps to balance the high salt content of the fish's ocean environment, thereby preventing it from drawing water out of the fish's cells, collapsing them. High concentrations of any compound would do the same thing, but too much of most compounds, including salt, would also damage the cell's vulnerable but vital proteins. High concentrations of TMAO do little or no harm to these and other proteins. TMAO belongs to a class of compounds called *compatible osmolytes*—those that provide osmotic protection without doing damage. TMAO also counteracts the inhibitory effects of the deep ocean's high hydrostatic pressure on certain of a fish's enzymes. TMAO attains particularly high concentrations in fishes in the polar regions, where it provides another benefit by acting as a sort of antifreeze. But no matter what climate they live in, all ocean fishes, including sharks, skates, and rays, as well as other marine animals such as octopus, squid, cuttlefish, and nautilus contain TMAO. They all begin to smell fishy when they die.

Only microbes can reduce odor-free trimethylamine oxide to fishy-smelling trimethylamine. So we know that when a fish starts to smell fishy, bacteria have begun to grow on it, although they probably will not do us any harm. But we are aware that the fish has been around long enough for bacteria to create anaerobic pockets and begin to produce highly odoriferous TMA. As fish lovers have long known, smelling is the best way to determine how fresh a fish is. Do it even if you have to poke a hole the plastic wrap at the supermarket. But don't bother with trout, catfish, tilapia, or carp. They are freshwater fish, so they don't contain TMAO and cannot support fishy-smelling anaerobic respiration. They can be far from fresh and smell pretty good.

The Black Sea or a Marine Mudflat

To find the next microbe, we could travel to the Black Sea or just look at a marine mudflat anywhere in the world. Because its deep waters are quite dark with a black cast, the Black Sea takes on a striking deep-blue color, although that is not the basis for its name, which is quite ancient and probably unrelated to color. The reason for the black cast of the depths of the Black Sea is microbial, as is the reason for the blackness and odor of mudflats that adjoin salt and brackish waters. This blackness and odor forms as a consequence of a kind of anaerobic respiration, in which bacteria derive metabolic energy using sulfate (SO_4^{2-}) as an oxygen substitute. In so doing they reduce sulfate to smelly hydrogen sulfide gas (H_2S). In other words, these sulfate-reducing bacteria convert SO_4^{2-} and organic nutrients into carbon dioxide (CO_2) and H_2S, releasing metabolic energy. Sulfate is abundant on Earth. Leached from land-masses, it accumulates in the oceans, but it is also found on land as its calcium salt, gypsum ($CaSO_4$), in layers that were once the bottoms of ancient seas. Gypsum is mined and used principally in building material. The white filler in drywall is gypsum, and so is blackboard chalk.

Hydrogen sulfide, the waste product of this form of anaerobic respiration, is a highly odoriferous compound, one we associate with rotten eggs. The reason for the association of hydrogen sulfide and eggs is also microbially based, though it does not depend on anaerobic respiration. The microbes that cause eggs to spoil produce this gas by splitting H_2S off sulfur-containing compounds that are present in eggs. This process forms only small amounts of hydrogen sulfide because eggs contain only relatively small amounts of these sulfur-containing compounds. But we are all too aware of such minuscule quantities of hydrogen sulfide because it has such an intense odor. Higher concentrations of hydrogen sulfide are extremely toxic. A single of breath of air containing as little as one percent hydrogen sulfide can cause loss of consciousness. However, hydrogen sulfide does not tend to accumulate in nature because it is so reactive.

In contrast to the very small amount of hydrogen sulfide microbes make from spoiling eggs, quite significant amounts are formed by the sulfate-dependent anaerobic respiration that occurs in the Black Sea and marine mudflats—a byproduct of the microbe's generation of metabolic energy. But why does this form of anaerobic respiration blacken the Black Sea and mudflats? Along with other compounds in the environment, hydrogen sulfide reacts with metal ions such as magnesium and manganese, forming insoluble metal sulfides, most of which are black. These small particles of black metal sulfides remain suspended in the water and mix with the sediment at the bottom of the Black Sea.

Because the hydrogen sulfide produced by sulfate-reducing bacteria is so reactive, it can also be destructive. For example, it reacts with metals such as iron, causing them to corrode. So pipes buried in mud exposed to marine water decay rapidly, exposed to the hydrogen sulfide produced by sulfate-reducing bacteria, which respire sulfate anaerobically. The most thoroughly studied representatives of sulfate-reducing bacteria are species of *Desulfovibrio*. These bacteria are totally committed to sulfate-dependent anaerobic respiration. They cannot use oxygen when it is available. They are even damaged by it.

We might ask what makes the Black Sea special among seas and oceans for its deep waters to be so noticeably blackened by sulfate-reducing bacteria. Like all oceans, the Black Sea contains sulfate,

organic nutrients, and sulfate-reducing bacteria. What makes it unusual is its extensive anaerobic environment. This intriguing body
of water has a huge oxygen-free zone because its lower level is saltier, colder, and therefore denser than its upper one. Being denser
than the waters above, the anaerobic region stays at the bottom.
The two layers do not mix. So the lower level never comes in contact with air and, as a result, it remains oxygen free.

Bodies of water, fresh or marine, with layers that do not mix
are termed *meromictic*. There are some meromictic lakes in North
Africa and the United States, notably Soap Lake in Washington
state as well as Green and Round lakes in New York state. They,
too, support populations of interesting anaerobic microbes in their
lower layers, though not sulfate-reducing bacteria. Fresh water usually contains only small amounts of sulfate, although salty desert
lakes are a notable exception. Most lakes are *holomictic*—they mix
once or twice a year, often in the spring when the lower layer is not
much colder and hence not much denser than the one above it.
Wind supplies the primary driving force. Most meromictic lakes
are protected from such wind action by being particularly deep and
having steep sides.

There are also many relatively small meromictic basins in the
oceans near the shore. They are maintained by a Black Sea–like halocline, a gradient of salt concentration. River water flows over
deep trenches of dense, salty water, not mixing. The Black Sea is by
far the largest of the world's meromictic basins, and it contains the
largest quantity of oxygen-free water on the planet. Its stabilizing
halocline is maintained by a constant flow of saline Mediterranean
water through the Bosporus Strait into its lower regions and a constant over-layering flow of fresh river water coming through the
Sea of Azov. The Mediterranean Sea is saltier even than other, larger
oceans because more water evaporates from it than flows into it as
fresh water.

Thus, sulfate-reducing bacteria are provided with a vast and ideal
venue in the lower levels of the Black Sea to convert extensive

amounts of sulfate to hydrogen sulfide. The sulfide combines there with metal ions, which are present in all marine environments, to form dark particles, black sulfides, giving the sea its characteristic and spectacular dark hue.

A similar set of conditions exists in most marine and estuarial mudflats. Seawater supplies sulfate. Organic nutrients and sulfate-reducing bacteria can be found in the mud, and just below its surface conditions are anaerobic because oxygen is rapidly depleted right at the surface by microbes that utilize oxygen for aerobic respiration. As a consequence, hydrogen sulfide is produced. Here again, the hydrogen sulfide reacts with metal ions in the mud, conferring its blackened color. The small amount of hydrogen sulfide that does not react with metal ions provides the rotten-egg odor characteristic of these mudflats. In contrast to marine mudflats, those around fresh water are much lighter in color and usually do not have a rotten-egg odor. Again, that is because, unlike oceans, most bodies of fresh water contain only small amounts of sulfate.

Of course, the Black Sea and marine mudflats are not the only locations where you will encounter black mud and a sulfurous odor. Wherever you see this mix of color and odor, you will most probably have found the sulfate-reducing bacteria.

Bubbles Rising from a Quiet Pond

Pause for a few minutes beside any quiet pond with a mud bottom—not a recent rain puddle but an established pond—and you are almost certain to notice an occasional burst of bubbles rise to the surface. With almost equal certainty these bubbles will be composed of methane (CH_4), what we commonly call natural gas.

It is easy to make certain. Cap the end of a small funnel with a plastic tube closed with a pinch clamp, or just some sort of stopper, and immerse it in the pond to fill it with water. Next, hold the tube inverted with the rim just below the surface of the pond in the region where you saw the bubbles, and with a stick, disturb the bottom of the pond to release a burst of bubbles that will displace the water and fill the funnel with gas. Now lower the funnel until just the cap protrudes from the pond in order to increase the pressure on the gas. Open the cap while holding a lighted match over the opening. The escaping methane gas will burn with a blue flame. So you have proven that it is not just air trapped in the sediment at the bottom of the pond; it is methane, made down there by archaea living in the muddy sediment.

We can be certain that these methane makers are archaea because a group of them, the *methanogens* (which simply means "methane makers"), are the only organisms on the planet that are able to make this gas. They do so by reducing CO_2 at the expense of compounds, either hydrogen gas (H_2) or some small molecule, commonly acetate (CH_3-COO^-), made in the sediment by some other microbe, almost certainly a bacterium. It is an uncomplicated reaction. Hydrogen gas and carbon dioxide are converted to methane and metabolic energy, albeit only a dribble of energy. Nevertheless, methanogens are able to harness it and flourish at the expense of it. Microbes are masterful energy misers.

Of course, there are many other anaerobic environments that contain hydrogen (or another appropriate small molecule) and CO_2. Usually methanogens proliferate and produce methane in them as well. One such environment is the specialized first chamber of the stomach of ruminates (cud-chewing animals such as cattle and sheep). The methane produced by ruminants is belched out. Collectively belching cattle, large and numerous, are a prodigious source of methane, estimated to be 20,000 metric tons daily. Because methane is such a frighteningly effective greenhouse gas, this source of the gas constitutes a significant environmental concern.

Most, perhaps all, of us humans also house at least some methane-producing archaea in our intestines. And about a third of us host sufficient quantities of them to be cowlike producers of methane, although fortunately humans produce much less of it.

About 20 percent of such methane is absorbed into the blood and eventually exhaled. That is the delicate way that gastroenterologists determine whether or not a person is a significant producer of methane, by testing for methane in exhaled breath. The rest of the intestinal methane is passed as flatus. Our major intestinal gas, however, is microbially produced hydrogen (H_2). Methane producers simply convert some of it to methane as they do in the mud at the bottom of ponds. The clear distinction between humans who do and do not produce significant quantities of methane has attracted the attention of gastroenterologists as well as microbiologists. Methane production was once suspected to correlate with susceptibility to colon cancer, but that ominous connection has been eliminated by subsequent studies. Production does, however, correlate with incidence of diverticulitis as well as constipation, although the distinction between cause and consequence is not clear. As I discuss later in the book, a connection between being a methane producer and gaining weight appears likely. This relationship between microbes and humans is being intensively studied.

The anaerobic digesters in sewage disposal plants provide another methanogen-friendly environment, as do waste disposal landfills. In well-managed sewage and waste disposal facilities, escaping methane is collected and burned for useful purposes. It may generate electricity or heat incoming sewage, thereby speeding the treatment process. Small digesters of human and animal waste are utilized in some countries, notably China, to generate methane at home for cooking and heating.

Local governments in rural China—the home of about 70 percent of the country's population—encourage the building and use of these small methane-producing digesters. The tanks, which can be constructed out of bricks or concrete in a week or so and cost about eighty dollars, are decidedly low tech, but they work quite well. They dispose of waste while producing a useful product. The digesters hold around twelve hundred gallons and are fed with human and animal waste, mostly from pigs, along with other agricultural waste materials, such as grain stalks and weeds that bacteria and archaea, working in tandem, convert into methane. From these materials the bacteria produce as endproducts the starting materials that the methanogens need to make methane. Most such home installations provide sufficient methane to supply around 60 percent of a family's energy needs for cooking and heating. An added and major benefit is the liquid effluent from the tank. Unlike the noxious feed, the microbe-processed effluent is odorless. Other microbes have depleted the odoriferous components, using them as nutrients. The tank's liquid effluent is a nitrogen-rich, dark-colored slurry that can be used to fertilize crops, and for a variety of other purposes as diverse as rearing worms that can be fed to chickens and fish.

Influenced by the West's increased sensitivity to energy concerns, the rural harnessing of methane-producing archaea is now spreading to other nations, including Germany and the United States. In 2005 the State of New York approved a subsidy to encourage small and mid-sized farms to convert their manure into methane to be used for generating electricity. A farm capable of generating as little as one megawatt of energy per hour can qualify. It takes about twenty-five cows' manure to generate that much power. The side benefit is a substantial decrease in offensive odor.

Although collectively methanogens produce massive amounts of methane, they are not Earth's only source of this gas. Methane is also formed geochemically in the high-temperature, high-pressure environment deep below Earth's surface near the mantle, the layer beneath Earth's crust. The methane made there also has a microbe connection. It serves as a source of nutrients that support the populations of microbes found several miles below Earth's surface. Not a place you might expect to find any living thing, but microbes thrive there on geochemical methane.

Indeed, methane is astoundingly abundant on this planet. Prodigious quantities of it occur in a solid form known as *hydrated methane*, a substance that was discovered only a few decades ago.

Hydrated methane (also called methane ice, methane hydrate, or methane clathrate) is a curious substance that looks like ice, but if a match is touched to it, it burns. In a sense hydrated methane is indeed a form of ice. It consists of water molecules arranged in an ice-like structure that traps molecules of methane within it. Although the methane content of the structure varies with pressure, it is always surprisingly high—up to 164 times the volume of the hydrate itself. Hydrated methane occurs mainly in two regions on Earth: the deep oceans near continental shelves, where it can be found in layers a thousand feet thick, and in the permafrost of the polar regions. Hydrated methane forms and remains stable only in environments where temperature is low, pressure is high, or both conditions exist. More carbon, it is estimated, is present on Earth in the form of hydrated methane than as petroleum. Hydrated methane also occurs elsewhere in our solar system—on the moons of Saturn, for example.

Hydrated methane, so far at least, has defied being harvested for use as fuel. Collectively, the two extensive reservoirs of microbially produced methane constitute a ticking environmental time bomb. At somewhat higher temperatures than now exist in our oceans or at a lesser pressure, hydrated methane spontaneously releases methane gas. Permafrost also releases methane as it thaws. The potential disaster-loop scenario is all too clear: increased temperatures resulting from global warming cause methane gas from hydrated methane to be released into the atmosphere, which increases the greenhouse effect, causing further warming, thereby releasing more methane, and on and on. Some earth scientists speculate that such a scenario actually played out 55 million years ago, causing the Paleocene-Eocene thermal maximum (a period when Earth was quite warm). This feedback loop holds the potential for a particularly rapid progression of warming because methane is such a potent greenhouse gas. Molecule for molecule methane is about fifty times more powerful than carbon dioxide. Even now with atmospheric concentrations of methane being two-hundredfold less than those of carbon dioxide, it is a significant contributor to the rapid global warming we are experiencing.

Once formed, methane is not microbiologically inert. Certain bacteria metabolize it via aerobic respiration in aerobic environments and others can use it as an energy source in the absence of oxygen. But once methane has escaped into the atmosphere, it is out of the reach of microbial methane users.

The innocent-looking bubbles rising from a quiet pond have multiple practical implications. Collected, they can serve as a fuel. Free in the atmosphere they contribute significantly, perhaps disastrously, to global warming, or as some say, catastrophic climate change.

Large Bubbles Rising from the Rancho La Brea Tar Pits

The Rancho La Brea Tar Pits (a translational redundancy and descriptional error because *la brea* means "tar," but they are composed of asphalt, not tar) are located near downtown Los Angeles, California. These pits are the residuals of pools of petroleum that began to accumulate some 40,000 years ago, quite recently on the geological scale of things and well within the experience of our own species, as upwellings from underground oil fields, forming a natural refinery as the crude oil reached the surface. More volatile gasolinelike components were released into the atmosphere, and residual asphalt was left behind in the deep pits formed by the upwelling. Some upwelling and distillation continues today: the area smells faintly like a gas station. It is in these areas where the large bubbles appear. In other parts of La Brea, the asphalt has hardened. Such regions have been mined for the ancient bones they contain. Surface asphalt pits are unusual, but they do occur elsewhere in the world. Similar but smaller asphalt pits are found in Bakersfield, California, and in Peru and Iran.

The pits at La Brea are the most well known and fascinating of all the asphalt pits because of their size and their yield of such a rich treasure of the preserved bones of animals entrapped in them over the centuries of their existence. During the last ice age, some 40,000 years ago, the spreading layers of asphalt covered perhaps by a layer of dust or leaves became a deceptively camouflaged, highly deadly trap for roaming animals. Alarm cries of trapped herbivores attracted carnivores, which also became entrapped. Asphalt-ensnared animals disappeared into the pit where their bones were preserved—a veritable animal census of a then cooler, damper, and presumably smog-free Los Angeles. It must be a highly biased census, however, because carnivore remains in the pits are about seven times more abundant than herbivore remains, quite the reverse of their distribution in most ecosystems. Typically, herbivores outnumber carnivores by a factor of around a hundred. But of course, entrapment was not random. Herbivores, presumably, wandered into it by unlucky chance; carnivores were attracted to it by the herbivores' desperate cries. A single trapped herbivore, therefore, might cause the entrapment of many carnivores.

But the complete biological census that the pits have provided is certainly rich and extensive. Plants (including a redwood tree and a now-threatened coastal live oak), mammals, birds (including a California condor, an eagle, and a turkey), reptiles (including a rattlesnake), amphibians, fish (including a rainbow trout), and invertebrates (including a scorpion and a termite) are represented. Many of the mammalian species, such as the American mastodon, the saber-toothed cat, the American lion, the American camel, the dire wolf, the American cheetah, and species of sloth, are now extinct. Fine specimens of some of them are on display in the George C. Page museum at the site. Other mammalian species, such as the American bison, cougar, bobcat, jaguar, llama, coyote, elk, raccoon, and skunk are still very much in evidence locally or elsewhere. Even the skeletal remains of one human, a 9,000-year-old Native American woman, were found in the pits.

The large bubbles, which rise from the still fluid-filled parts of the pits, are filled with methane, a gas frequently associated with oil fields as well as quiet ponds. It was tacitly assumed that the source of the methane in the bubbles was the underground oil field and that its origin was ancient and perhaps geochemical. But a recent microbial census of the pits conducted by Jong-Shik Kim and David E. Crowley of the University of California, Riverside, changed that.

A bubbly tar pit seems like a particularly unlikely and inhospitable home for microbes. The pits are loaded with toxic chemicals and precious little water. But the survey, which analyzed a chunk of asphalt taken from twenty feet below the surface (to minimize the contribution of recent arrivals) showed that the pits teem with microbes, including 200 to 300 previously unknown species of bacteria. On the basis of its location in the pit, the sample was estimated to be about 14,000 years old. Part of this census survey was microbiologically conventional. Samples of asphalt were extracted with water, which was then used as source material to start cultures by standard methods of laboratory cultivation. A few microbes were recovered this way. Most of the microbes in the pits could not be cultured. So how did we know they were there?

Most microbes, probably somewhere around 99 percent of them, cannot yet be cultivated in the laboratory. The reasons for their stubborn defiance of cultivation are varied, complex, and probably, to a certain extent, still unknown. Certainly some of the reasons include the mutual interdependence of microbial strains and species (and therefore their inability to survive a solitary laboratory existence), sensitivity to higher than just minimal concentrations of required nutrients, and possibly extremely slow growth (thereby outlasting the microbiologist's patience). When microbiologists realized just how many more microbes existed than they were able to cultivate in the lab, it came as somewhat of a shock. Certainly, only a very minimal cultivation gap was perceived during the golden age of discovery in medical microbiology, when the microbes that caused many infectious diseases were studied by laboratory cultivation. Only a few of these disease-causing microbes, such as the one that causes leprosy, resisted cultivation in the laboratory. That riots of vintners, disgruntled by others using and profiting from their good name. Now the name is legally protected; a similar wine made elsewhere is a sparkling wine. In tribute to the long tradition and high quality of these particular French wines, almost all sparkling wines are put in the same slope-shouldered bottles, even though vintners may no longer call them "champagne."

The wired, bulging cork used as a closure and the heft of the bottle are more than paeans to tradition, however. They are microbe-imposed necessities, needed to contain the pressure of the wine's carbonation. If such serious containment measures are needed for sparkling wines, we might ask why other carbonated drinks from cola to club soda are kept in ordinary bottles or cans. Most carbonated drinks are charged with carbon dioxide gas only to a pressure that the can or bottle can withstand. Sparkling wines are charged by yeasts that produce carbon dioxide and alcohol through alcoholic fermentation of grape sugar until building pressure stops further production. The yeast, not the bottler, determines the pressure of carbon dioxide within the bottle. This yeast-set pressure turns out to be about eight atmospheres (118 pounds per square inch) of pressure, and a heavy glass bottle with a wired cork closure can contain such a pressure. Maynard Amerine, father of the California wine industry, liked to say that this happy coincidence of biological and physical fact was the palpable proof that God loves us.

When champagne and high-quality sparkling wines from other grape-growing regions are made by the traditional *méthode champenoise*, carbonation is generated in the same bottle in which the final product is marketed. To each heavy bottle is added a finished still wine and a *liqueur de triage* before it is capped. The *triage* is a small amount of a mixture of sugar and yeast cells. By converting the sugar to alcohol and carbon dioxide through fermentation, the yeast carbonates the wine. Winemakers like to say that they calculate the amount of sugar needed to reach the ideal pressure in the bottle. But, in fact, the yeast makes this decision.

By traditional methods, the bottles are stored in an A-frame rack that holds the bottles in a slanted position with necks down as this

string of spectacular successes gave microbiologists the perhaps smug assurance that all microbes could be cultured in the laboratory. It was just a test of the microbiologist's skills. But these disease-causing microbes are, of course, a special case. They evolved to flourish by themselves (in "pure culture") in the laboratory of our bodies.

More elaborate and sophisticated methods of laboratory culture are constantly being developed, but for the moment, at least, a high-tech end run is being made on the problem. Rather than trying to cultivate the microbe in the lab, researchers recover and sequence DNA from its environment. On the basis of DNA similarities to microbes that have already been studied in the laboratory, they deduce the uncultivated microbe's closest relatives and hence what it is doing in nature. This approach has come to be known as *metagenomics*. It has produced a lot of new information about microbes, but it depends on the reservoir of knowledge gained through laboratory cultivation. Sooner or later, that reservoir will have to be replenished by traditional methods of painstaking cultivation and careful study.

Metagenomics was used to make the microbial census of the tar pit. Samples of asphalt were frozen with liquid nitrogen, then DNA was extracted, sequenced, and analyzed. It told a wonderful new microbial story. Certain bacteria in the pits attack hydrocarbons of the asphalt and convert them to more readily usable compounds, including fatty acids. Methanogenic archaea convert the fatty acids into the methane that causes the basketball-sized bubbles to emerge from the pits. Of course, it is a circumstantial story, but it is a very convincing one.

The huge bubbles rising from a tar pit are indeed a microbe sighting, albeit one only recently established because understanding depended on newly acquired technology. It is likely that the La Brea tar pit bubbles portend microbial discoveries elsewhere. Microbial ecology, as we will see, is undergoing a near-explosive growth phase.

3

Food and Drink

The connections between microbes and food are typically microbophobic: after all, they spoil our food and cause foodborne illnesses, some serious, even lethal. But on the positive side of the ledger, from our earliest history, we have depended on microbes to do quite the opposite—to produce food and preserve it. We have used microbes this way long before we knew they existed. Only recently have we come to understand the complexity of microbes' role in these ancient processes. Once you have sampled this rich reservoir of interactions between microbes and food, you will undoubtedly be able to make and explore many other fascinating microbe sightings in your kitchen and on your dinner table.

Much of microbially based food making and preservation is dependent on *fermentation*, a familiar term that describes yet another scheme by which some microbes acquire metabolic energy. The two schemes discussed in previous chapters, aerobic and anaerobic respiration, exploit energy-yielding oxidation/reduction reactions. So does fermentation, but in a different way. As Louis Pasteur, modern microbiology's acknowledged founding father, pointed out in a much-quoted, slightly incomplete statement when he discovered the scheme, "Fermentation is the consequence of life without air."

He didn't know about anaerobic respiration. Respirations exploit an oxidation of organic nutrients by reducing oxygen or an oxygen substitute. In a fermentation, the organic nutrient oxidizes and reduces itself. One part of the nutrient molecule becomes oxidized, the other reduced. (Rarely, an organic nutrient oxidizes a different organic nutrient.) In some fermentations, the sugar (the organic nutrient) is split, yielding one endproduct that is more highly oxidized and one that is more reduced. Such is the case with alcoholic fermentation: glucose is converted into CO_2, ethanol (C_2H_6O), and metabolic energy. Carbon dioxide is more oxidized than glucose (its oxygen to carbon ratio is higher and its hydrogen to carbon ratio is lower), and ethanol is more reduced than glucose (its oxygen to carbon ratio is lower and its hydrogen to carbon ratio is higher). In other fermentations, such as the lactic acid fermentation, one part of the same product molecule (lactic acid) becomes more oxidized and another part becomes more reduced.

All fermentations, regardless of type, share one important characteristic: they do not yield much energy. So microbes that employ them must process large quantities of organic nutrients in order to generate sufficient metabolic energy for growth. As a consequence they produce a large quantity of endproducts, which accumulate in their environment and change it in ways that are readily apparent to us.

A Bottle of Champagne

A bottle of champagne contains a full sensory experience for a microbe watcher. The special look, sound, taste, and feel of the bubbles are all the result of microbial activity. Of course, properly and legally, a beverage named "champagne" can only be produced in the Champagne district of Burgundy. It is an *Appellation d'origin contrôlée* ("controlled term of origin"), in fact the one for which the system was instituted in the early twentieth century to quell pressure-generating fermentation proceeds and for four or five weeks after that. Daily, the bottles are given about a quarter turn, causing the clumped yeast cells that settle on the lower side of the bottle to work their way down towards its neck, a process called "riddling" or *remuage*. The riddler always wears a protective face shield because bottles do occasionally explode from the pressure produced by the fermenting yeasts.

Now many champagne makers use automatic riddling machines, so that after the triage is added, the bottles are arranged upside down in shipping cases, which are placed on a platform that vibrates gently for an hour four times a day and is rocked every two minutes. Of course, there is little agreement as to whether traditional or automated riddling gives the better product. Those that employ cost-saving automated riddling make a virtue of their break with tradition by claiming their method increases precision and yields a more consistent product.

During the long contact between yeast and wine that occurs during riddling, some yeast cells autolyze (break open by the action of their own enzymes), spilling their contents into the wine and enriching its flavor. Most experts agree that sparkling wines made by the alternative Charmat method, in which carbonation occurs in large pressurized tanks from which bottles are filled under pressure, does not have equivalent quality to those made traditionally. One reason is probably the lesser amount of autolysis that occurs in the tank.

In the *méthode champenoise*, when the clump of yeast reaches the neck as a consequence of riddling, either traditional or automated, the contents in the end of the bottle are frozen, by dipping the neck in brine at 0°F, forming a plug that is forced out when the bottle is held upright and the temporary plug is removed. The yeast cells are disgorged along with a small amount of champagne. Then the lost volume is made up in the process of "dosage" by adding an equivalent amount of *liqueur d'expédition,* usually sweetened still wine.

A microbial fermentation that is stopped by only eight atmospheres of pressure is extraordinary. Under most conditions it takes pressures in the range of 250 to 500 atmospheres (3,700 to 7,400 pounds per square inch) to stop alcoholic fermentation.

Why a wine fermentation in a pressurized container is especially sensitive to elevated pressure, stopping at the relatively modest, bottle-containable pressure of eight atmospheres, is not well understood, although the general principles of the effect of elevated pressures on microbes are. Most microbiologists who have studied the phenomenon of champagne agree that it is a complex interaction between pressure, concentration of alcohol, and the amount of dissolved carbon dioxide that is present. Whatever the reason, it makes champagne and other sparkling wines possible.

The phenomenon has also been exploited in the making of some still wines using methods developed in Germany. The primary fermentation of the crushed grapes (called *must* by winemakers) takes place in a pressurized steel tank. By regulating the pressure in the tank, the winemaker can maintain a slow and steady rate of fermentation. The rate-limiting pressure need not be very high. Maintaining a fermenting yeast's self-generated pressure at 0.6 atmospheres (about 9 pounds per square inch, which is not even enough to inflate an automobile tire) is sufficient to decrease the yeast's rate of growth fourfold. Such pressure regulation avoids the potentially damaging consequence of a too-rapid fermentation: excessive heat that can damage the quality of the finished wine or even injure the yeast cells sufficiently to stop the fermentation, causing a dreaded "stuck" fermentation. The method was evaluated but never adopted in California.

Holes in Swiss Cheese

We have been making cheese for a very long time—perhaps as long as 10,000 years, when humans settled down and began to live in groups. But until very recently, our silent partner in the cheese-making process went unacknowledged. Cheese is a microbial production from beginning to end, and microbes account for cheese's rich variations in style, taste, and appearance, including the eyes in Swiss cheeses. This human-microbe partnership is highly productive. Worldwide about 13 million tons of cheese are made annually.

Even the production of cheese's raw material, milk, depends on microbes' unique ability to break down cellulose. And microbes, with or without our help, intervene immediately in making milk into cheese. Milk is largely a solution and suspension of sugar (lactose), protein (for the most part casein), and fat.

The most ubiquitous microbes that participate in cheese making are a mixture of lactic acid bacteria, usually including *Lactococcus lactis,* though many other species of this group of bacteria also contribute to producing particular styles of cheese. For a number of reasons, lactic acid bacteria are ideal for the purpose. They are exclusive fermenters. That is all they do. Although they tolerate being exposed to air, they do not use oxygen and, therefore, do not oxidatively damage any of milk's subtle flavor components. And as Louis Pasteur discovered just in 1857, they convert sugars, including milk's lactose, into lactic acid. (Some species, grouped as heterofermentative lactic acid bacteria, do produce other compounds, also by fermentation, along with lactic acid.) And lactic acid bacteria are always present in fresh milk, probably entering it from the cow's udder, ready to go. The lactic acid they produce makes two vital contributions to cheese making: because lactic acid is a strong acid, its powerful acidity protects the milk from being spoiled by other microbes, and it prepares the milk for the next step of the process: curd formation.

Traditionally, cheese making depended on the lactic acid bacteria that were naturally present in the cheese-making facility. Of course, over time, populations of certain kinds of lactic acid bacteria became established in certain locations on the equipment and in the buildings—so cheese makers there could depend on year-to-year consistency. Some became famous for it. Certain types of cheeses could only be produced in certain places. But that is not the case now. Once location-specific traditional cheeses are now made in various countries worldwide. For example, Camembert cheese, once made only in its namesake village in the Normandy district of France, is now made in prize-winning quality in such distant places as Ireland and Petaluma, California. To accomplish this cheese-making ubiquity, starter cultures of microbes are added to initiate the process. Such cultures must be carefully selected, a process requiring detailed genetic information about lactic acid bacteria that has been acquired only relatively recently. With intense research, an impressive package of information is now available. To begin with, the complete chromosomal genome sequence—base by base—has been determined for over twenty-five lactic acid bacteria. Each one has 2 to 3 million base pairs to be lined up. But even the complete chromosomal genome sequence does not tell the whole story. Most lactic acid bacteria carry several small genetic elements called *plasmids*. Although plasmids do not encode essential metabolic activities, the lactic acid fermentation for example, they do encode reactions that are vital to making a quality cheese. To be a good starter culture for a particular cheese, the bacterium must carry the proper array of plasmids.

The starter culture grows vigorously, reaching populations as crowded as a billion cells per milliliter. Consequently, the milk becomes quite acid and ready for the next step, curd formation. To form a curd, casein must be precipitated from the milk. This is traditionally accomplished by the nonmicrobial process of adding rennet. Rennet is an extract of the dried lining of the fourth chamber (abomasum) of a suckling cow's or another ruminant's stomach. Rennin, the active ingredient of rennet, is an enzyme (chymosin) that cleaves milk's casein at a precise location into two protein fragments. The smaller one (sixty-three amino acids long), like casein, is soluble, becoming part of the whey, the watery substance left behind after the casein precipitates out of the milk. The larger protein fragment (163 amino acids long) is insoluble and aggregates in chunks called "curds," the stuff from which cheese is made. This clipping is the first and necessary step in casein digestion. Dedicated vegetarians consider rennet-made cheese a breach of their principles, so they have resorted to other methods. Vegetarian cheese is available through the benevolence of microbes. Some fungi naturally make chymosin-like enzymes, and genes from cattle have been introduced into certain fungi (the yeast, *Kluyveromyces lactis,* and the filamentous fungus, *Aspergillus niger*) as well as *E. coli* so that they, too, become sources of rennin.

We might ask how people 10,000 years ago gained the seemingly obscure knowledge that they needed an extract of a suckling calf's stomach to make cheese. A reasonable guess is that someone used a ruminant's stomach (a goat's would be a reasonable size) as a convenient tote bag for milk. The requirements for starting to make cheese were all there—lactic acid bacteria from the milk and rennet from the tote bag. At the end of the journey, the milk-delivery person had fresh cheese.

Today, as we have seen, the march of the microbes has even overtaken this tote-bag step of cheese making. Some chymosin is obtained from the fungi that make it, but most chymosin now used in cheese making is obtained from microbes with a chymosin-encoding gene from a ruminant added by recombinant DNA technology. The curds and whey are then separated, a process usually facilitated by heating and sometimes, depending on the type of cheese, by pressing. Whey is a voluminous leftover. Now that family pigs are no longer readily available to consume it, it has become a waste disposal challenge.

Next the separated curds are cut into uniformly sized pieces, the size of which again is determined by the type of cheese to be made. The curd is formed into the cheese's characteristic shape, and the highly complex microbe-mediated process of maturation begins. That is the stage of cheese making when characteristic and subtle flavors are developed. Bacteria do most of the job, but in some

cheeses—notably Camembert, Roquefort, and Stilton—certain
fungi play vital roles. Maturation involves a number of microbe-
mediated chemical processes—some brought about by intact microbes and others by enzymes that spill out of microbial cells as
they autolyze. Most of these conversions are not too well understood, but high on the list is breaking the split casein into smaller
pieces, sometimes as small as individual amino acids, some of
which, glutamate and aspartate for example, add their own characteristic flavors. And certain amino acids are further changed by the
action of microbes into other flavor components. It is a complex
endeavor.

Diacetyl is one of the important ingredients that microbes add.
This simple four-carbon compound adds a buttery flavor that en-
riches the taste of cheese and also butter. Even artificial butters use
it for that reason. But it is not so agreeable in beer; in fact, its presence in beer is a major defect. You have probably noticed it, particularly in a slow-moving draft beer. The easy way to detect diacetyl is to pour a small amount of beer onto the palm of your
hand and rub it till the liquid evaporates. The residual odor on your
hands is distinctive. Some brewers describe it as a "foxy" odor. It
smells like a fox's den—with no modern-day sexual connotation.

Because of its buttery flavor, diacetyl is added to a variety of
foods, including microwave popcorn, a use that has caused dozens
of workers to develop a debilitating lung disease known as "popcorn workers' lung." Legislation is pending that would set standards for occupational exposure to diacetyl. But I suspect the informative beer test is safe.

The holes or "eyes" as well as the distinctive flavor of Swiss-type
cheeses are also microbially based. Traditionally, the bacterium
Propionibacterium shermanii, which makes Swiss cheese be Swiss
cheese, just happened to be present in the cheese-making facility.
Now pure cultures of it are added early in the cheese-making process along with the lactic acid bacteria. The genus has another species, *Propionibacterium acnes,* that is associated with acne, causing
irritation by converting some of the excess sebum produced by the

skin's sebaceous glands into irritating fatty acids. They cause the angry pimples, which in severe cases can become disfiguring.

Like lactic acid bacteria, species of *Propionibacterium* obtain their metabolic energy by fermentation. They oxidize some of the lactic acid that the lactic acid bacteria make to carbon dioxide, yielding acetic acid as a byproduct while reducing other molecules of it to propionic acid.

The carbon dioxide makes the holes. Propionic acid and other metabolic products of *Propionibacterium* give Swiss cheese its sweet, nutty flavor. Strains of *Propionibacterium* are carefully chosen to give the proper amounts of these products. Even the amount of carbon dioxide is critically important. There must be enough to form the holes—if not it is called "blind" Swiss cheese—but not an amount that would split the cheese.

The products of the propionic acid fermentation do more than just add flavor. Propionic acid is also an effective and nontoxic preservative. It (in the form of calcium propionate) has long been added to bread to prevent the growth of bread mold. The mixture of the products of a propionic acid fermentation are also effective. Such natural preservatives are made by cultivating *P. shermanii* in skim milk and drying the culture as a powder. These products, with trade names such as MicroGARD, are added to cottage cheese and other foods.

The holes in Swiss cheese are a rather specific example of the microbe *Propionibacterium shermanii*. But many other microbes make their contribution to this and all cheeses. Microbes are full partners in making quality cheeses. The enterprise cannot proceed without them.

A Bottle of Vinho Verde

Vinho verde, a wine from the northwestern corner of Portugal where wine presumably like this has been made since the twelfth century, sometimes comes in a distinctive and attractive bulbous bottle. Its contents constitute an especially intriguing and scientifically instructive example of microbes at work.

"Vinho verde" means green wine in Portuguese, though the wine in the bottle can be red, rosé, or, as is usually the case, white. The *verde* probably refers to the wine's age, because it is usually consumed the same year that it is made. Its alcohol content is low, around 7 percent, which is about half of most table wines from California. It is also semi-sparkling or *pétillant*. That is why it qualifies as a special microbe sighting.

The wine's semi-sparkling characteristic is caused by carbon dioxide, low concentrations of it, but its origin is quite different from that in sparkling wines such as Champagne. For one thing, bacteria, not yeast, make it; and for another the source of the carbon dioxide is an organic acid—malic acid, not sugar. (Grapes contain two organic acids, malic and tartaric, along with lesser amounts of citric acid and two sugars, glucose and fructose.) Many lactic acid bacteria (those capable of fermenting sugar to produce lactic acid) can convert malic acid ($C_4H_6O_5$) into lactic acid ($C_3H_6O_3$) and carbon dioxide (CO_2), releasing metabolic energy, even in the presence of wine's low pH and elevated alcohol content, a process called *malolactic fermentation.*

But this metabolic conversion is not really a fermentation. No oxidation and reduction occurs. The malic acid (HOOC-CH_2-CHOH-COOH) is just split into two pieces: one fragment is carbon dioxide (CO_2) and the other is lactic acid (CH_3-CHOH-COOH). This molecular cleavage event releases a small dab of energy, which malolactic bacteria are able to capture and utilize as their sole source of metabolic energy. We will consider how these microbes carry off this impressive metabolic feat later in this section.

The consequences of the malolactic fermentation for the wine are threefold: it lessens the wine's acidity (because malic acid has two acidity-conferring carboxyl groups [-COOH] and lactic acid has only one); it carbonates the wine (because CO_2 is produced), and it increases the wine's flavor complexity (malolactic bacteria add small amounts of flavor components to the wine).

Pétillant wines, such as most vinhos verdes, are highly unusual, but the malolactic acid fermentation itself is quite common. It occurs in almost all red wines. In fact, red wines from cooler grape-growing regions such as Europe (including France) would be too acidic to be high-quality wines were it not for the ameliorating intervention of the malolactic fermentation.

The malolactic fermentation was traditionally one of enology's (wine making's) most romantic mysteries, both favored and dreaded, a free-spirited microbial act of nature that occurred when, where, and with what results it seemed to choose. In California, for example, up until about 1960, most vintners did not even know whether their wines had undergone a malolactic fermentation. Some were aware of it when several months after vintage, large storage tanks (mostly made of redwood in those days) would rumble gently, announcing that the malolactic acid fermentation was under way. Vintners did not start it. They could not stop it. It just happened.

Using rapid analytical methods that had become available about that time, I, then an assistant professor in the Department of Enology at the University of California, Davis, found that most California wines did undergo a malolactic fermentation and that the particular malolactic bacterium that brought about the fermentation varied from winery to winery, but not from year to year. A particular malolactic bacterium must set up permanent residence in a winery, its hiding place most probably the holding or fermentation tanks. The redwood tanks, with their porous surfaces, offered copious hiding places where bacteria could persist from one vintage to the next.

At the time, there were mixed enological emotions about the malolactic fermentation. Most vintners were somewhat embarrassed that beyond their control, bacteria in their winery were attacking their wine and changing it. Bacterial activity in a winery was traditionally associated with spoilage and lack of cleanliness.

After all, the great Louis Pasteur in his famous *Études sur le vinaigre et sur le vin* (studies of vinegar and wine) had analyzed and classified the bacterial "maladies of wine," many of which were caused by lactic acid bacteria. On the other hand, malolactic fermentation was known to occur in great European red wines. It was certain to decrease acidity and thought to increase flavor complexity. And, moreover, no one wanted the fermentation to occur in the bottle. The resulting bacterial cells would cloud the wine and the fermentation would frequently cause a buttery flavor considered unpleasant, in some cases more unpleasant than others depending on which strain or species of bacteria was the culprit. (The strain of malolactic bacteria that adds a bit of sparkle to a bottle of vinho verde adds only a slight pleasant flavor.) If the fermentation occurred in a holding tank well before bottling, the undesirable aspects of the flavor dissipated before bottling. Then the bottled wine was more reliably stable.

Starting a malolactic fermentation with a favorable bacterium at an appropriate time became a goal of wine making. But it proved to be elusive. Adding bacteria obtained from a good wine that had undergone a malolactic fermentation to a fresh one did not start a fermentation. The reasons proved to be issues of microbial nutrition, both of yeast and malolactic bacteria.

Yeasts have undemanding nutritional needs. They can make all vital nutrients that microbiologists call *growth factors*, such as amino acids and vitamins, so they do not need a supply from their environment. But yeasts are notoriously greedy nutritional misers. After they have finished growing, they scavenge nutrients, including vital ones, from their environment and store them within intracellular sacks called *vacuoles*.

This microbial avarice had an interesting historical side effect on human health in the 1930s. It was popular then for young people, encouraged by the manufacturers of baker's yeast, to consume small yeast cakes (composed of live cells) as a presumed preventative or cure for acne. In some cases, the practice resulted in vitamin deficiency because the live yeast cells stole essential nutrients from human intestines.

Malolactic bacteria, like lactic acid bacteria in general, are quite the opposite from yeasts. They do not hoard nutrients, but their complements of biosynthetic enzymes are so limited that they need a considerable package of growth factors in order to grow. The lactic acid bacterium *Leuconostoc citrovorum*, for example, must be supplied nineteen amino acids (nine more than we need) and ten vitamins to satisfy its growth needs.

The nutritional conflict between yeast and malolactic bacteria explains the difficulty of initiating a malolactic fermentation at will. After the initial yeast fermentation is complete, yeasts gorge themselves on the growth factors in the new wine. By the time the alcoholic fermentation carried out by yeasts (the so-called *primary fermentation*) is complete, not enough growth factors remain in the wine to support the growth of malolactic bacteria. Growth of these bacteria and their attendant malolactic fermentation cannot occur until several months later, after sufficient quantities of yeast cells settling on the bottom of the tank have autolyzed, releasing growth factors in adequate amounts to support growth of the malolactic bacteria resident in the winery.

The solution to the problem of how to start a malolactic fermentation seemed to be the addition of a healthy inoculum of a pure culture of malolactic bacteria to the wine while the primary yeast fermentation was still under way, before the yeast cells could nutritionally impoverish the wine. The scheme worked. Using this scheme and a strain of malolactic bacteria, ML34 (for the thirty-fourth strain of malolactic bacteria), that I had isolated from the Louis Martini Winery in the Napa Valley, the winemaker Brad Webb and I successfully started the first induced malolactic fermentation at the Hanzell Winery in the Sonoma Valley. Later the strain was assigned to a new species and given a proper Linnean binomial, *Oenococcus oenis*. It was made freely available and used worldwide. Starting and managing the malolactic fermentation is now a standard tool of wine making.

But a profound and intriguing scientific question remained, which was addressed and answered by others, namely, how do malolactic bacteria glean sufficient energy from the malolactic acid fermentation to satisfy their need for metabolic energy? The reaction that changes malic acid into lactic acid and CO_2 releases only a little over a third as much energy as is needed to make a single molecule of ATP, the indivisible unit currency of metabolic energy. Malolactic acid bacteria, which might be the record holders among prokaryotic microbes' many notable stingy energy gleaners, save that dab of energy, and when they have accumulated enough of them, they make a molecule of ATP.

They do this by means of an ATP-generating method called *chemiosmosis,* which all organisms employ in one form or another. In chemiosmosis a proton gradient (a difference in the concentration of hydrogen ions) is established across a membrane; in the case of malolactic bacteria and other prokaryotes, the cytoplasmic membrane. In nature, any difference in concentration of any substance tends to neutralize itself, and the only way a trans-membrane proton gradient can neutralize itself is for the higher concentration of hydrogen ions outside the cell to flow through the cytoplasmic membrane back into the cell. The only route is through a trans-membrane pore made of an enzyme called ATP synthase. When hydrogen ions flow through this enzyme pore (and they are forced to by the gradient), ATP is generated. Perhaps the enzyme could be viewed as being like a water wheel, generating mechanical energy from a flow of water.

The unique aspect of the malolactic fermentation is how the proton gradient is generated. It is done just by shuffling malic acid into the cell while expelling lactic acid and carbon dioxide (Figure 3.1). One hydrogen ion enters with the molecule of malic acid and one leaves with each of the two products of fermentation, lactic acid and CO_2, thus creating a one-ion gradient with each molecule of malic acid utilized. The dab of energy that the reaction releases drives the shuffling. It takes three protons flowing through the ATP synthase to generate a single molecule of ATP. So protons accumulating outside the cell act as a bank account in which metabolic energy is saved. When enough has been accumulated, it is converted into the cell's energy currency, ATP. As we noted, nature offers no free lunches, even very meager ones, but small savings assiduously accumulated can produce wealth. A few microbes vary the theme of chemiosmosis by using a gradient of sodium rather than hydrogen ions. But the principles are the same in both cases: the established gradient of ions drives a synthesis of ATP as the ions flow back into the cell through a membrane-spanning enzyme.

Malolactic fermentation is a generous benefactor. Winemakers and wine drinkers benefit from a less acidic, tastier red wine. Vinho verde consumers can enjoy its light, sparkling appeal. And *Oenococcus oenis* satisfies all its needs for metabolic energy. This

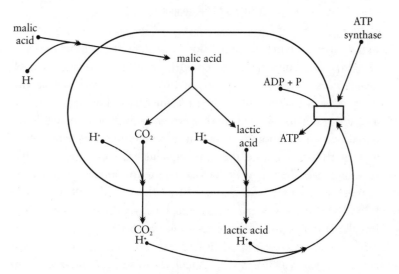

Gleaning metabolic energy via malolactic fermentation.

type of fermentation also occurs during the manufacture of pick-
les, but it is hard to make a positive case for its role there. The carbon dioxide it produces occasionally becomes trapped inside some
fermenting cucumbers, swelling them. These swollen cucumbers,
called "bloaters," have to be thrown out. Control of bloater formation is symptomatic, not preventative: the pH of the pickle vat is
lowered to release bound carbon dioxide and air or nitrogen is
bubbled through the vat to remove the carbon dioxide before
enough has accumulated bloat the pickle.

The production of vinho verde provides insight into how microbes can grow at the expense of reactions that yield only infinitesimal amounts of energy. Think of it that way and tell your dinner companions next time you have a bottle of it. Possibly, the
malolactic bacteria such as those in vinho verde are record holders in the category of living successfully by exploiting low energy-
producing reactions.

A Bottle of Château d'Yquem

In the rating of wines from the Bordeaux district of France in 1855
requested by Emperor Napoleon III, wineries were assigned to
groups, designated first, second, or third *crus* (growths). Château
d'Yquem was assigned to the first (Premier Cru), but only it among
all Bordeaux wines was given the further distinction of being designated Supérieur. Thomas Jefferson, a noted appreciator of wines,
apparently had come to the same conclusion. He visited the cha-
teau in 1784 when he was the U.S. envoy to France. He liked the
wine so much that he ordered 250 bottles for himself and additional bottles for George Washington. Let us hope they cost him
less than today's price per bottle—well over $150.

The microbe associated with Château d'Yquem is a fungus, *Botrytis cinerea*, which has a spectacularly mixed reputation. In addition to its patrician contribution to Château d'Yquem (and of

course related wines), it causes an unsightly, potentially lethal gray-mold blight of herbaceous plants, a disease that is particularly prevalent in cool, damp climates. *B. cinerea* also infects bunches of grapes late in the growing season, causing bunch rot, or gray mold. Most grape growers are anxious to avoid this blight because it can cause the bunches to ooze their contents, encouraging the growth of other fungi and insects. *B. cinerea* is, however, also a vital contributor to making some of the world's most admired dessert wines, including Château d'Yquem. In this role it goes by the decidedly more upscale name of "noble rot."

In many respects *Botrytis cinerea* is a rather ordinary and typical fungus. It grows, as most fungi do, as a mycelium, that complex of intertwining protoplasm-filled tubes that are not subdivided into cells. *B. cinerea* is an example of an imperfect fungus that cannot find a partner. Those strains that do find one turn out to be ascomycetes and are given the name *Botryotinia fuckeliana* (sorry about that).

When *B. cinerea* grows on a bunch of grapes, it nourishes itself by forming bulbous structures (called *haustoria*) on the tips of branches of its mycelium. These press close against the grape's cell membrane, drawing out some of the berry's water and nutrients through it (much as the fungus in a lichen draws nutrients from its phototroph partner). In doing so, it passes a few of its own constituents back to the berry. This predominantly one-way exchange is selective and just what a winemaker wants. Taking out water concentrates the grapes' juices, making them much sweeter. And *B. cinerea* does not detract from the sweetness by using the grape's sugars (an equal mixture of glucose and fructose as you will recall). Instead it utilizes only the grape's malic acid for its own metabolism. That also benefits the winemaker by decreasing the grapes' acidity and softening the wine to be made from it, rather like malolactic fermentation does in a bottle of vinho verde. *B. cinerea* also gives Château d'Yquem its rich golden color and its distinctive subtle flavor, undoubtedly a complex mixture of many compounds.

And *B. cinerea* makes a further contribution to the wine. It adds glycerin, which gives Sauternes wines their remarkable mouth-filling feeling of thickness.

Other microbes also make special contributions to Sauternes wines. The naturally occurring yeasts that are present on these grapes (a mixture of Semillon and Chenin Blanc in the Sauternes district of France) have a preference for glucose, leaving predominantly the much-sweeter fructose in the finished wine. Thus through combined gifts of two microbes, *Botrytis cinerea* and the resident wine yeast, Sauternes wines are sweet, golden-colored dessert wines, with a distinctive flavor and mouth-filling richness.

Sauternes are the prize of the botryticized wines and Château d'Yquem is the prize of Sauternes. France is not the only producer of such wines, however. Hungarian Tokaj is also highly admired, as are many German wines. California produces very few because of its drier and, therefore, less hospitable climate for *Botrytis*. A good year and climate for *Botrytis* is one in which a prolonged humid spell (to initiate and spread the *Botrytis* infection) is followed by a drier period (to desiccate the infected grapes). Germany's botryticized wines are named and classified by when and how the grapes are harvested. In increasing *Botrytis* involvement and therefore quality: Spätlese (taken late) means harvest of the grapes has been delayed, giving more time for *Botrytis* to become established; Auslese (selected) means that *Botrytis*-infected bunches have been selected; Beerenauslese (selected berries) means that infected berries have been selected; and Trokenbeerenauslese (selected dry berries) means that individual dry berries have been selected. Of course, expense tracks with quality.

Château d'Yquem is a microbe sighting recognizable by almost all of our senses: visually by its golden color, by taste as well as by odor, and by feel of its glycerol-conferred thickness. Curiously its microbiological connections are all natural events. Winemakers do not consciously add any of these microbes. In all respects, the transition to our next sighting, a bottle of cola, is a great leap.

A Bottle of Cola

This comparatively mundane microbe sighting is made simply by examining the list of ingredients on the labels of soft drink bottles. Recorded in relative abundance, high-fructose corn syrup (HFCS) comes right after carbonated water. And HFCS is not restricted to carbonated drinks. A brief survey of supermarket labels shows it is everywhere in prepared foods that are sweet or slightly sweet, with the exception of "diet" foods, from fruit juices to salad dressings, even ketchup. For such uses, HFCS has largely replaced sugar (sucrose) because HFCS is very sweet and cheap. We continue to use more and more of it. Already, more HFCS than sucrose is used worldwide. Over 10 million tons of HFCS are produced annually. It is consumed principally in the United States, but overseas use is growing rapidly.

Corn, except for fresh sweet corn, is not sweet at all. Corn syrup —the sort you might find on your kitchen shelf—is moderately sweet. HFCS is very sweet. Both corn sweeteners are made using enzymes obtained from microbes. Any stroll through a grocery store will yield a host of other products that use enzymes as well. Laundry detergents, digestive aids for persons with lactose intolerance, preparations for suppressing flatulence after consuming beans, and processes from making chocolate candies with soft, cherry-filled centers to clarifying wines and beer also depend on microbial enzymes. The use of enzymes from microbes is pervasive.

Particular microbes are cultivated in huge vessels called *fermenters,* and enzymes are extracted from them. The variety of enzymes and their uses are vast. Considering just a smattering of them gives an impression of their rich diversity.

One of the most intriguing and certainly compelling for candy lovers is the use of the enzyme sucrase to make those wonderful chocolate-covered cherry candies with liquid centers. Coating a liquid with melted chocolate sounds difficult or impossible. But that is not the way it is done. When the centers are dipped in the chocolate, they are solid because of the considerable amount of sucrose they contain. The microbial enzyme sucrase, also called *invertase,* is added to the mix. Slowly, after the center is coated with the melted chocolate and it has solidified, sucrase converts the relatively insoluble sucrose into a mixture of much more soluble glucose and fructose, thereby converting the solid center into a liquid one.

Lactase, which is also obtained from yeast, is used as a digestive aid for people with lactose intolerance; they suffer gastric distress when they consume milk products because they are unable to digest lactose, the sugar mammalian milk and most milk products contain. Sufferers are unable to make their own lactase, which cleaves lactose into its component sugars, galactose and glucose, which humans can digest.

As infants, almost all humans can produce lactase, which they must have to digest their mother's milk or formula made of cow's milk. But members of many ethnic groups lose that ability as they mature. (Northern Europeans and certain Africans are notable exceptions.) Those that lose the ability to make lactase become lactose intolerant. Uncleaved lactose, which cannot be absorbed in the small intestine, passes into the large intestine, causing abdominal pain, diarrhea, bloating, and flatulence. With a dietary supplement of microbe-derived lactase, those who are lactose intolerant can consume milk, ice cream, cheese, yogurt, and other dairy products without subsequent intestinal distress.

Microbe-derived enzymes are also used as a dietary supplement to relieve a sugar intolerance that all humans share, the inability to digest certain multi-component sugars (with wonderful names like raffinose, stachyose, and inulin). When we consume vegetable foods, notably beans that contain significant quantities of such sugars (chemists more accurately would say oligosaccharides), they pass undigested through our stomach and small intestines into the large intestines. There, bacteria attack them and produce gas—hydrogen and for some of us methane. Then we and those around us suffer from flatulence. The principal microbe-derived enzyme in gas-reducing remedies is alpha-galactosidase, derived from lactic acid bacteria such as *Lactobacillus plantarum* and others. These break down the two or three flatulence-inducing sugars that are linked together into their component sugars (including glucose, galactose, and raffinose). These are then absorbed in our small intestines before they can enter the large intestines to be utilized by the gas-producing microbes that reside there.

Microbe-derived enzymes are also used in huge amounts in laundry detergents to increase their cleaning power. Microbe-derived lipase (from *Aspergillus niger* and related fungi), which cleaves water-insoluble fats and oils into soluble constituents (glycerol and fatty acids), removes grease stains. Proteases break down proteins, especially those from the bacteria belonging to the genus *Bacillus*. *B. subtilis* and *B. licheniformis* solubilize proteins, including bloodstains, so that they can be washed away.

And microbe-derived enzymes are also used to beautify prepared foods. Pectinase, from *Aspergillus niger* and other fungi, breaks down the plant-wall substance pectin. In food production, it is used to clarify fruit juices and wine. Beta-gluconase, from *B. subtilis,* which breaks down another plant-wall substance, removes the haze from beer.

Although all of these enzymes have an impact on what we eat, drink, and use to clean our clothes, the microbe-derived enzymes that are used to make corn into sweeteners have probably had the greatest impact on human affairs. Industrially, the starch in corn is converted into HFCS in three steps: starch becomes maltodextrans (soluble fragments of starch composed of long chains of glucose), which becomes glucose, then HFCS. Each of these is catalyzed by a microbe-derived enzyme. Enzymes that mediate the first two steps have been used for a long time. And glucose, the product of their combined actions, is indeed a sugar, but not a very sweet one. That product when made from cornstarch is corn syrup, which is used largely in cooking. The huge jolt in sweetening power occurs in the

third step, when glucose is converted into fructose. Fructose is very
sweet. If table sugar, sucrose, were assigned a sweetness value of
100, glucose would be only 70, but fructose would be 130. Moreover, glucose is relatively insoluble, so it tends to precipitate from
concentrated syrups, making them difficult to store and, therefore,
to use in the modern food-processing industry. Fructose does not
present this problem. It is twice as soluble as glucose. So, converting about half the glucose in a syrup to fructose (producing HFCS)
solves both problems: the product is as sweet as sucrose and it is a
stable solution that doesn't readily precipitate if chilled.

The microbial enzyme glucose isomerase, which converts glucose
into fructose, is a curious one. In retrospect, its discovery seems
highly improbable. The metabolic role it plays in the bacteria that
produce it has nothing to do with glucose or fructose. Instead it
catalyzes the first step in the utilization of another sugar, xylose,
which is abundant in nature because it is a constituent of hemicellulose, a major component of the cell walls of plants. The enzyme
converts xylose to another sugar, xyulose, which the bacteria can
then utilize. In 1957, quite by accident, the enzyme was discovered
to have a feeble capacity (about 160-fold less than its ability to act
on xylose) to convert glucose into fructose. This discovery was not
completely surprising because glucose is structurally similar to xylose. Soon thereafter another improbable, fortuitous, but very important discovery revealed that the enzyme became a much better
glucose-to-fructose converter in the presence of just a little cobalt
ion. This discovery opened a period of intensive research on the
glucose isomerase enzyme, which is widely distributed among bacteria, that led to its practical use in producing HFCS.

The research advanced on several fronts. Microbes were treated
with mutagens (mutation-inducing chemicals or radiation) and,
among the resulting mutant strains, those that produced greater
glucose-to-fructose converting capacities were selected. And methods of cultivating the producing microbe were developed that
caused them to yield more enzyme. Among the many hurdles that
were cleared in order for production of HFCS to become commer-

cially feasible was eliminating the necessity of adding expensive xylose during cultivation. Without xylose, naturally occurring microbial strains make no enzyme. The presence of xylose signals (induces) the microbe to make the enzyme that metabolizes it. This economic barrier was breached when mutant strains (called *constitutive strains*) were found that make glucose isomerase even in the complete absence of expensive xylose. Also, more effective means of using the enzyme were developed. The most important of these was attaching the enzyme chemically to inert material packed into a long column. Then, rather than having to add fresh enzyme to each batch of glucose-containing syrup, a number of batches could be passed through the same column. That technical advance offered several advantages: less of this relatively expensive enzyme was needed, and the required cobalt ion could also be immobilized on the column, thereby keeping it out of the product.

But one major technical challenge remains unmet to this day. The fructose sweetener made this way must be used as a syrup. It cannot be made to crystallize, so it has no granular form, like sucrose. Crystallization fails because the glucose isomerase–catalyzed reaction does not convert all the glucose to fructose. It produces only an equilibrium mixture of the two sugars because the reaction proceeds in both directions: glucose to fructose and almost as rapidly back to glucose. Sugars in solution must be relatively pure in order to crystallize. The equilibrium mixture does not yield a sufficiently pure solution of fructose. This sounds like a completely chemical dilemma, but its resolution might well prove to be microbial. The equilibrium between glucose and fructose shifts toward fructose as temperature is increased. If glucose isomerase could tolerate a temperature high enough, it might be able to produce a crystallizable solution of fructose. But where is such a microbe to be found? In a high temperature environment? In a mutant strain that produces a heat-tolerant form of the enzyme? The answer, yet unknown, will surely be lucrative.

In many respects HFCS seems to be an ideal way to satisfy our near insatiable sweet tooth. It is very sweet. It is inexpensive. And although almost all artificial sweeteners are medically suspect for one reason or another, HFCS must be safe. Fructose (fruit sugar) is the major natural sweetener of most fruits. And most important, HFCS does not contain sucrose. Sucrose is the cause of dental caries. Eliminating sucrose completely from our diets would eliminate cavities. But unfortunately, extensive use of HFCS has brought its own set of unintended negative consequences.

The incorporation of massive amounts of HFCS into our collective diets corresponded with the onset of the obesity epidemic and an associated dramatic rise in juvenile diabetes. Some see a direct causal connection. There is more evidence for the association between HFCS and the obesity epidemic than mere coincidence. Owing to its modest price, the sizes of soft drinks served in fast food establishments and restaurants have increased severalfold. And there seems to be a palpable physiological explanation for its obesity-causing tendencies. In response to glucose, the sugar we derive from most of the food we ingest, the pancreas produces insulin, which induces the release of leptin, the hormone that signals satiety. Insufficient leptin inevitably leads to obesity: for example, mice strains that are genetically incapable of making leptin are obese. But the pancreas has no receptor for fructose. So fructose does not initiate the cascade that tells us we have had enough to eat.

But let us not blame the microbe. It was only trying to eke out a living on the breakdown products of materials in the cell walls of plants. We used it to get fat. On the other hand, as we will see in another chapter, certain other microbes might well contribute directly to obesity.

An Unrefrigerated Egg

Raw eggs in the shell are remarkably free of microbes. They spoil slowly because their whites have a set of proteins—avidin, lysozyme, and conalbumin—with powerful abilities to stop microbial growth and even to kill gram-positive bacteria. Raw eggs can last for months without spoiling, although their quality and ability to fluff up an angel food cake or make creamy mayonnaise does decline for nonmicrobial reasons (largely the loss of CO_2, which gives egg white its stiffness). In contrast to the microbe-resistant raw egg, an intact hardboiled egg, in which the defensive proteins have been inactivated by heat, spoils rather rapidly—within a week or so. The hardboiled egg's susceptibility to spoilage shows that major protection from microbes does not depend primarily on the egg's shell and membrane, though it does offer some. A cooked egg out of the shell spoils even more rapidly.

Avidin is a protein that protects by binding to the vitamin biotin, making it unavailable to those microbes that, like us, must have it. Biotin-requiring microbes cannot grow in its absence, so they are unable to proliferate in an egg because biotin-binding avidin reduces the concentration of the available vitamin to extremely low concentrations. In addition to egg whites, avidin is found in the tissues of birds, reptiles, and amphibians, where it also fulfills an antimicrobial role. Avidin binds to biotin more tenaciously than any other protein is known to bind to any other small molecule. This extraordinary avidity for biotin has been exploited in a number of biochemical procedures, for example, the purification of proteins. It is only necessary to attach a molecule of biotin to the desired protein. Then, immobilized avidin can fish minuscule amounts of that particular protein in pure form out of a complex mixture of other proteins, leaving the rest of them behind.

Lysozyme, the egg's second line of antimicrobial defense, is a bit more widespread in nature than avidin. It is found in tears, nasal secretions, saliva, and the intestinal mucosa as well as egg white. Lysozyme has a remarkable bacterium-killing specificity: it catalyzes the cleavage of certain bonds in peptidoglycan, the macromolecule that comprises the cell wall of bacteria. When these bonds are broken, the bacterium's wall is weakened and pressure on the cell wall from within causes the cell to explode, that is, fall apart or lyse. Curiously, lysozyme kills bacteria in almost the same way that penicillin does, by attacking their cell walls. It is also curious that both lysozyme and penicillin were discovered by the same microbiologist, the Scotsman Alexander Fleming (1881–1955). He discovered lysozyme in 1921 after observing that his own nasal mucus killed certain bacteria. He called the active ingredient, which turned out to be an enzyme, *lysozyme* because it killed bacteria by lysing them. A year later Fleming reported finding lysozyme in tears, saliva, and egg white. Fleming's interest in lysozyme was for use as a possible antimicrobial chemotherapeutic agent with which to treat bacterial infections. But it proved not to be effective. Lysozyme is, however, given credit for stimulating Fleming's interest in chemotherapy, which in 1928 led to his momentous discovery of penicillin. Fleming's other stimulating influence came from having served as an army surgeon in World War I and witnessing, first hand, the cruel lethality of wound infections.

With only a few exceptions, all bacteria have walls composed of peptidoglycan, but not all bacteria are killed by lysozyme. Lysozyme must, of course, have access to peptidoglycan in order to destroy it, and the outer membrane of gram-negative bacteria denies such access. So the lethality of lysozyme is restricted to gram-positive bacteria.

Although the growth-restrictive action of avidin is limited to biotin-requiring microbes and the lethality of lysozyme to gram-positive bacteria, conalbumin, the egg's third line of antimicrobial defense, is equal opportunity. It acts on virtually all microbes. Conalbumin (sometimes called *ovotransferrin* because of its similarity to transferrin, an iron-transport protein found in mammals) binds tightly to iron (Fe^{3+}), making it unavailable to invading microbes. With only a few possible exceptions, all microbes, indeed all cellular organisms, need iron for certain of their vital constituents. They cannot grow without iron. A dramatic proof of conalbumin's vital role in protecting eggs from microbial spoilage occurred in the past when eggs were washed commercially in iron trays. Such eggs spoiled rapidly because enough iron had entered the egg to overwhelm the iron-binding capacity of its conalbumin, leaving some iron available to spoilage microbes.

Lack of available iron in our blood (there, too, it is largely sequestered by an iron-binding protein, transferrin) also affords us some protection against microbial infection. For this reason, some pediatricians advise against supplementing infants' diets with iron unless they show signs of anemia.

The egg is a success story of targeted chemical defenses against microbes. Its protective proteins afford an effective, nontoxic way to preserve food.

Thick Italian Salad Dressing

Italian salad dressing in a jar at the grocery store used to be watery; now it is surprisingly thick. You might wonder why. The label provides an answer: "contains xanthan gum." So do many other foods, including other dressings, sauces, frozen foods, and beverages, even toothpaste. Xanthan gum is also used in drilling fluids called "mud" —the viscous liquid that is pumped by the bits (cutting surfaces) of drilling rigs in oil wells and thereby forced back to the surface, lifting pieces of rock and earth with it. All these uses of xanthan gum depend on its unusual viscosity-conferring properties. As little as a half-percent and sometimes only a tenth that much (0.05%) is sufficient to thicken a liquid such as salad dressing. Moreover, the thickened liquids exhibit the curious phenomenon of pseudoplasticity: they thin when subjected to shearing (twisting force), a property that is essential for making thickened foods such as salad dressings pleasantly palatable. Chewing and other movement of your mouth subject foods to shearing, so if they are thickened with xanthan gum, they do not feel heavy in the mouth. They are thick when poured, but thin as they are moved about the mouth.

Xanthan gum is not the only extensively used thickening agent. Among others, various forms of starch, alginic acid (from kelp), carrageenan (from the red alga, or Irish moss), and agar (from another red alga) are widespread. But because of its unusual properties, xanthan gum is gaining increasing popularity. So it may come as a bit of a surprise to hear that xanthan gum is composed of bacterial corpses.

The particular dead bacterial cells that comprise xanthan gum are those of *Xanthomonas campestris,* a bacterium that infects plants. Unlike previous sightings, this one is not the result of the metabolic activities of *X. campestris.* Here we encounter the microbial cells themselves, particularly the outer part of them. Many microbes, including fungi, archaea, and bacteria such as *X. campestris,* surround themselves with an extracellular layer of slime called a *capsule,* which is sometimes many times larger than the cell itself.

Usually the capsule is composed of polysaccharide, though occasionally it is protein. The one around *X. campestris* is polysaccha-

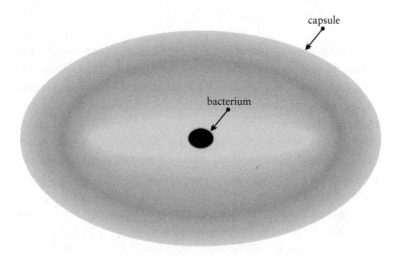

A bacterium's capsule sometimes dwarfs the cell.

ride. A capsule benefits the microbe it surrounds in a number of ways: for example, the capsule keeps the microbe from drying if it becomes beached out of its usual liquid environment. Also the capsule can glue the microbe to a solid surface, thereby preventing its cells from being swept away from a favorable location. In certain cases, the capsule is associated with infection because it allows the microbe to cause disease. Indeed, the pathogenicity (disease-causing ability) of some microbes is completely dependent on their capsules. That is the case with *X. campestris*. But for unknown reasons, *X. campestris* forms extraordinary amounts of capsule—much more than we would expect is needed to cause an infection. I will return to the connection between capsule and pathogenicity later.

To make xanthan gum, one simply grows *X. campestris* in a fermenter (an industrial term that ignores the inconvenient fact that fermentation is an anaerobic process and the contents of the tank are highly aerobic). To maintain the aerobic conditions that *X. campestris* requires, air is pumped rapidly into the fermenter and the contents are stirred vigorously. The results are astounding. Because the microbial cells produce such huge quantities of capsule, the entire culture becomes so thick that it pours only extremely slowly out of an open beaker. Precipitating the capsule along with entrapped bacterial cells and killing them by adding an organic solvent, then drying the solid material, yields the light-colored powder that is xanthan gum.

Owing to its infection-enhancing powers, the lowly, slimy microbial capsule has had a huge impact on human affairs. Possibly the most notorious capsule is the one that surrounds *Streptococcus pneumoniae,* commonly called the *pneumococcus,* the bacterium that causes a serious and, if untreated, lethal infection of humans, pneumococcal pneumonia. This dreadful disease, now readily controlled by penicillin and vaccination, was a scourge of humanity well into the early twentieth century, even in wealthier countries such as the United States. Healthy adults in the prime of life would suddenly sicken, suffer from high, unrelenting fever and progressively severe respiratory problems, and die within days. The long list of the disease's fatal victims includes presidents (William Henry Harrison), authors (Leo Tolstoy), and generals (Stonewall Jackson).

Microbiologists have long known that the pneumococcus's ability to cause deadly disease is totally dependent on its capsule. Phagocytes, those white blood cells that engulf and destroy most invading bacterial cells, are unable to engulf pneumococcal cells. Their capsules prevent it.

But not all untreated pneumococcal infections end in death. Over half of those infected recover. If they survive for six to ten days, their immune systems spectacularly come to their rescue. It takes that long for the immune system to begin making the antibodies (called *opsonins*) against the capsule that allow phagocytes to engulf and destroy the pneumococcal cells. The patient then undergoes the dramatic "crisis" often depicted in novels and film. The flickering candle by the bedside burns brightly again. Drenching sweat and a sudden drop of the patient's temperature signal the beginning of complete recovery. The attending doctor wipes his brow and smiles.

Mutant strains without capsules are completely harmless, causing no disease whatsoever. This dramatic dependence of pneumococcus on its capsule, a life-or-death distinction for us, especially intrigued a British microbiologist, Frederick Griffith, and led him in 1928 to perform a simple experiment that led to the greatest discovery in biology of the twentieth century. Griffith identified deoxyribonucleic acid (DNA) as the molecule from which genes are made, the molecule in which all life's instructions and accumulated evolutionary knowledge are stored.

Griffith wondered if the polysaccharide that composed the capsule itself rendered pneumococci virulent (capable of infecting and causing disease), or if it must actually surround the pneumococcal cell to exert its lethal effect. In other words, would a mutant strain of pneumococcus that did not have a capsule (termed *rough* because of the appearance of its colonies on plates) and therefore unable to cause disease become virulent if it were mixed with capsular material from a "smooth" (capsule-forming and therefore virulent) strain? For feasibility, rather than using pure capsule for his studies, he decided to use heat-killed cells of a smooth strain.

He injected mice with a mixture of live cells of a rough strain and heat-killed cells of a smooth strain. The mice died, but control mice injected only with either one of the components of the mixture remained quite healthy.

The experiment seemed to provide a clear and unequivocal answer: the capsule need not surround a cell to make it virulent; capsular material just had to be present. But Griffith was suspicious of this seemingly obvious and clear-cut result. He pursued the issue further by isolating pneumococci from the lungs of the dead mice that had been injected with the mixture of the cells from the rough strain and the heat-killed smooth strain. The results astounded him. The bacteria he found in the lungs were smooth: a capsule surrounded them. Did that mean that some of the injected smooth cells had survived exposure to heat? Further experiments showed that this was not the case. He went on to characterize in greater detail the smooth cells he had isolated from the lungs of the dead mice and found that they had certain properties in common with the dead smooth cells that he had injected into the mice, as well as other properties in common with the live rough cells. The surviving cells were apparently hybrids of the two strains he had injected.

Then Griffith made his great intellectual leap. He speculated that something had come out of the dead smooth cells and had "transformed" (genetically altered) some of the rough cells, conferring on them the ability to make a capsule and giving them other smooth-cell properties as well. It was these genetically altered cells that proliferated and killed the mice. Griffith called this mysterious something from the dead cells the "transforming principle."

Further experiments soon established that the mice were irrelevant to transformation. Extracts from one strain of pneumococcus could be added to another strain in a test tube and change or transform its characteristics in a variety of ways. And these changes "bred true." The bacteria and their progeny did not revert to their original state. They were indeed genetically altered, not temporarily modified by the presence of the extract. The "transforming principle" appeared to act as though it were a solution of genes that could be taken from one strain of pneumococcus and, when added to another, would incorporate into them and as a result confer a new genetic makeup. It seems odd that a cell can take up genes floating around in its environment, but some species of bacteria can do exactly that. This form of genetic exchange, which occurs naturally among many kinds of microbes and can be induced to occur artificially between other kinds of cultured cells, is now called *transformation.*

A team of microbiologists at Rockefeller University, Oswald Avery, Colin Macleod, and Maclyn McCarty, recognized the significance of Griffith's results and proceeded to obtain the transforming principle in pure form so they could analyze it. They realized that if they learned what the transforming principle was made of, they would know what genes are made of. To the surprise of almost all, and the reluctant acceptance by many scientists, Avery and his colleagues found the answer in the relatively simple four-monomer macromolecule, DNA. Conventional wisdom had held that genes were so complicated that they must be composed of the much more complicated twenty-monomer macromolecule, protein. But as we all know now, genes are made of DNA. Griffith's experiments and astute observations on a microbe illuminated a path to this momentous discovery.

The bacterial cell's capsule looks insignificant. It is just a gooey layer that surrounds some bacterial cells. But it has become a major article of commerce as a gelling agent useful in both the food and the mining industries. Study of it has also led to the discovery that DNA encodes the genetic information of all forms of life, an important milestone in the history of biological science.

4

Living Together

Living together is a characteristic pattern of microbial life. Very few microbes live in nature as isolated, individual cells, although that is the way they are usually studied in the laboratory. Many communal microbes live imbedded in masses of other microbial cells like themselves, in aggregates called *biofilms*. Frequently such masses become arrestingly colorful, as in the case of the hot springs at Yellowstone National Park. In this chapter we will sample the vast number of microbes that live in close association with some other quite different form of life, often to the benefit of both partners.

Spanish Moss

Spanish moss hangs in mood-setting festoons from trees in the southeastern United States and farther south, all the way to Argentina. Neither originating in Spain nor a moss, it is as beautiful as it is badly named. It is a flowering plant belonging to the bromeliads, the family that also includes pineapples. Similar festoons appear on trees in the West and other parts of the country, but although they are also called Spanish moss, they are not even plants.

They are visible aggregations of two kinds of microbial cells: a phototroph (one capable of photosynthesis, either a green alga or a cyanobacterium) and a fungus, living together intimately for their own and the other's benefit, a living arrangement called a *mutualistic symbiosis*. The pair goes by a name that might be more familiar—a lichen. The association is so intimate and the appearance of each pair is so distinctive that biologists name lichens as though they were individual organisms rather than pairs of them. The great eighteenth-century naturalist Carolus Linnaeus, who introduced the logical system of naming organisms that we still use today, started it all. He and his last student, Erik Acharius, who specialized in the study of these intriguing microbial associations, named them as species of a single genus, *Lichen*. Present-day students of these symbioses have carried the concept of classifying these pairs of organisms as though they were a single species even further. Now, lichens are grouped into genera and families within a phylum assigned to the fungi. Individual associations are still assigned Latin binomials according to the Linnaean system. The Spanish moss–like lichen, for example, is named *Ramalina menziesii*.

Linnaeus and his students believed that lichens were individual species, but it has been known since 1867 when they were studied by Simon Schwendener that that is not the case. He demonstrated convincingly that each individual lichen is a combination of at least two distinct organisms. If further proof were needed, the two partners can be teased apart and cultivated separately. Growing alone, they look like ordinary fungi and algae or cyanobacteria. But when they are mixed together again under the right conditions, they form a lichen that has the characteristic structure (a thalus) of the original lichen from which the individual species were obtained.

Not only does each lichen look like a distinct species, they can reproduce vegetatively as such. On their surfaces, lichens produce small, dustlike particles called *soredia,* which consist of one or several cells of the photosynthetic partner surrounded by the fungus's hyphae, collectively making up its mycelium. These micro-lichen packages can be transported great distances by wind or water. If they come to rest in a suitable location, each one can develop into a new lichen thallus.

Each of the partners that makes up a lichen lives not only in a symbiotic relationship, but a mutualistic one. Each contributes to the shared welfare and benefits of their association. That is what mutualism means. Symbiosis alone is a value-neutral term with no implications about which partner benefits, is harmed (as with parasitism), or is completely unaffected by the other (as with commensalism). Symbiosis simply means living together.

The phototroph is the lichen's food provider. By photosynthesis it makes sufficient carbohydrate food to satisfy both partners. The fungus provides the housing structure and some food as well. It supplies a few mineral nutrients that its phototrophic partner needs by attacking and dissolving a bit of the surface that the lichen is clinging to, be it rock or some other substance.

A few lichens are a *ménage à trois,* composed of a fungus and two different phototrophs, usually a green alga and a cyanobacterium. There is no known explanation for the necessity or even the advantage of three microbes living together in this way. But there must be one because when particular trios do live together, they form a distinct species-like association.

Some lichenologists—a special breed of microbiologists with their own International Association, yearly International Congress, and published list of endangered lichen species—suspect that the fungal-phototroph relationship is a bit more sinister than mere symbiosis. They look at it as a form of parasitism or perhaps more accurately slavery perpetrated by the fungus. In this view, the fungus captures the phototroph and uses it for its own purposes, keeping it alive and healthy for its own benefit. A certain amount of physical evidence supports such a concept. Most fungi in lichens produce specialized structures (haustoria) that are pressed tightly against the phototroph, occasionally even penetrating it, presumably to steal its contents. Other lichenologists put it a bit more gently, characterizing the phototrophs as vegetative cattle, tended by a benign fungal herdsman.

In certain cases, especially those flat lichens that grow on rocks, the dual nature of a lichen can be observed directly. The outer ring of the thalus, usually identifiable by its different color, sometimes white and sometimes black, is a photosynthesis-free zone where the fungus is outgrowing and thereby extending beyond its photosynthetic partner. This zone is pure fungus; the central region is the normal fungal-algal or fungal-cyanobacterial combination of a lichen.

Whatever the motivation of the participants in the relationship, lichens are an extremely successful evolutionary experiment. They occur in a vast variety of shapes, sizes, and habitats. There are about 13,500 recognized species. Some thrive where no other organisms can, for example on the surface of granite rock in a dry climate. It is estimated that lichens are the dominant vegetation of about 8 percent of Earth's terrestrial surface, a significant competitor with grassland, for example, which is dominant on about 40 percent of its surface.

Quite a variety of organisms participate in the lichen lifestyle. Both major groups of perfect fungi (Ascomycota and Basidiomycota) can enter into it, and a staggering fraction of them do. Approximately 42 percent of ascomycetes are lichen formers. Lesser numbers and a lesser percentage of basidiomycetes participate, as can also be said for cyanobacteria. Indeed, more than 95 percent of all species of lichens consist of an ascomycete in partnership with a green alga. Twenty-five genera of green algae form lichen partnerships, and fifteen cyanobacterial genera do. But lichen-component microbes are not totally committed to living as lichens. Most are also found in nature as free-living creatures.

One of the principal reasons for lichens' success is their ability to feed on air. They obtain everything they need—moisture as well as all other nutrients—from the atmosphere, except for the mineral nutrients that the fungal partner extracts from the surface to which the lichen clings. Being air feeders, one would think that lichen's preferred photosynthetic partner would be a cyanobacterium because most cyanobacteria are nitrogen fixers. They are able to supply the partnership with usable nitrogen from atmospheric nitrogen gas (N_2) as well as organic nutrients from carbon dioxide.

But surprisingly, as we have noted, cyanobacteria are not the fungus's preferred partner. This indifference to the seemingly enormous selective advantage of having a partner that can supply fixed nitrogen as a nutrient might reflect a fundamental fact of lichen life: they grow extremely slowly, so slowly that the abundant usable nitrogen the cyanobacterium could supply would not benefit a lichen to any significant extent. A lichen can apparently get all the nitrogen it needs to prosper by utilizing the minute amount of ammonia that is always present in the atmosphere, perhaps augmenting it with an occasional, tiny supplemental dose of nitrogen that might be splashed onto the thalus from another source—nearby nitrogen-rich soil, for example. Indeed, the cyanobacteria that are found in lichens form very few of the specialized cells (heterocysts) they use for fixing nitrogen. Presumably the lichen's need for fixed nitrogen is so modest that it demands very little from its nitrogen provider, and therefore does not signal it to produce more nitrogen-fixing cells.

The case is a bit different for the tripartite lichens (those that contain a fungus, a green alga, and a cyanobacterium). The cyanobacteria they contain produce a large number of heterocysts, and they fix nitrogen vigorously. Atmospheric fixed nitrogen compounds such as ammonia are apparently inadequate to satisfy this more nitrogen-greedy group of lichens.

Most lichens do grow remarkably slowly. The crustose type (those that grow flat on a solid surface, such as a rock or a tree) is clearly the slowest. The foliose type (those that grow as a series of leaflike lobes) is a bit faster, and the fructicose (those that are attached to a solid surface at a single point and form complex branched structures) is slightly faster still. Slow does, for sure, mean slow. Many crustose species grow at rates less than a half a millimeter in diameter per year. So a small one of these, about an inch in diameter, would be over fifty years old.

Growth rates have been measured by yearly observations of the same lichen and by measuring the size of crustose lichens on objects of known age, such as gravestones, stones in mine tailings, or those in old stone buildings. From this sort of information it is possible to judge the age of particular lichen species in other places. Individual lichens in the Arctic, for example, are estimated to be extremely old, in the range of five thousand years. Numbers such as these suggest that lichens might well be the oldest living things on our planet. The oldest known tree, "Promesius," is a bristlecone pine *(Pinus longaeva)* near Wheeler Peak, Nevada, in Great Basin National Park. It contained 4,900 countable yearly growth rings when, tragically, it was cut down in 1964. The distinctly odoriferous creosote bush *(Larrea tridentate)* of the Sonoran Desert in the American Southwest is also a competitor for the longevity record and so are certain large fungi. The mighty Sierran sequoia, once a leader, lost out in the longevity competition some time ago.

Lichens' very slow growth rate—and knowing something about how slow it actually is—has opened an intriguing field called *lichenometry* that is being put to use for a number of purposes, including dating objects and estimating changing levels of pollution. Lichenometry reverses the procedure of measuring the size of lichens on objects of known age to determine their rate of growth and uses this growth-rate information to estimate age of an object, or more precisely the time its surface became available for lichens to grow there, by measuring the size of the lichens on it. The measurements are particularly valuable for estimating ages of objects up to about 500 years, an extremely useful range because it covers the awkward stretch that is too recent for radiocarbon dating to be very accurate. Such lichenometry dating has proved useful in estimating how long it has been since the rocks in glacial moraines were thrown up there by a passing glacier and how rapidly glaciers have been retreating, exposing rock surfaces for lichens to colonize. It was also used to establish that the frequency of avalanche activity has been on the increase, presumably as a consequence of global warming.

Lichens also record a history of air pollution. Recent observations on growth of lichens have shown that certain types of them in certain regions are growing more rapidly now than they did in the past, presumably because of increased air pollution, which adds to the air more nutrients that stimulate the rate of lichen growth. Such data are being used to map changing patterns of pollution.

Lichens make other positive contributions as well. They constitute a major food source for reindeer and caribou. Indeed, lichens constitute about 90 percent of the winter diet of caribou in the Arctic, even when it is covered with snow. Reindeer and caribou can smell lichens through the snow and dig down to get them. Occasionally there are fights over a good patch of lichens.

In British Columbia, Canada, there are two kinds of caribou, each of which lives on the particular kinds of lichens available in their habitat. Both live in the Cariboo Forest region. One kind, the mountain caribou, lives in the higher mountains of the region where the snows in the winter can be around fifteen feet deep. They have adapted to feed on arboreal lichens, those that grow on trees. Under such conditions, it is about the only forage available to them. The other kind of caribou, the northern caribou, live in a region that has been heavily logged. They have adapted to feed on terrestrial lichens.

The microbial symbiosis we know as lichens provides sustenance for many other creatures as well. Some say that manna from heaven described in the book of Exodus as sustaining the children of Israel when they were wandering in the wilderness must have been lichens that can flourish under such hostile conditions.

Spanish moss is just one of many lichens. Look for them, for example, on tombstones and compare their size to the inscribed dates on it, or those on a tile roof compared with the house's age. You can even compare the size of lichens on houses of known similar ages in order to speculate how well the houses are insulated.

A Belching Cow

The occasional belching of a cow contentedly eating grass in a field, chewing its cud, and growing fat on a diet that would not keep most animals alive, is a lifestyle made possible by microbes. Microbes in the cow's stomach cause it to belch, but they also enable cows to eat grass. As we will see, the strategy of housing microbes in the gut is key to the way that a number of other animals, including some insects, are able to prosper on a meager diet made up largely of cellulose, a plant substance that we and all other animals cannot digest on our own.

Belching, which cows do a lot, is a direct consequence of microbial metabolism. It is the necessary release of gas, principally methane, produced constantly as a byproduct of the microbial metabolism occurring in the cow's elaborate digestive system. A typical cow produces and *eructates* (the more refined scientific term) about seven and a half gallons of gas daily. If for any reason a cow becomes unable to belch out this gas, it is in serious trouble. The cow swells like a balloon, a condition called *bloat*. If untreated, bloat is lethal to the cow. Lancing, piercing the side of the cow with a long knife to release the constantly accumulating gas, is the veterinarian's heroic treatment for bloat. Surprisingly, perhaps, opening the rumen, the microbe-packed chamber of the cow's stomach, directly to the environment by lancing or other means does no apparent damage to the cow. In fact, researchers who study the rumen routinely keep a fistulated cow, one with a permanent covered transparent opening to the rumen, in order to take repeated samples from it, called by some of those in the business a rumen with a view. The cow remains healthy, seemingly still contented.

Chewing the cud also has a microbial connection. It is more than an expression of contentment. It is essential for the cow's survival. In order to nourish itself adequately, the cow must grind the grass and other plant material it consumes into particles fine enough for microbes in the rumen to attack the cellulose effectively. Cows ingest feed rapidly. Then, later, they regurgitate small boluses of feed (the cud) to rechew them into finer bits, a process called *rumination*. Cows ruminate about ten hours a day.

But how does a cow maintain itself and even fatten on a diet of grass or hay? Of course, grass or hay does contain small amounts of protein, sugars, and fats, nutrients that all animals, including us, are able to use as food. But there is not nearly enough of these components to fill an animal's nutritional needs. The major organic constituents of grass, hay, and other leafy plant materials are cellulose and hemicellulose, the fibrous structural materials comprising the walls of plant cells. These macromolecules are made up of strings of sugar molecules, all of which animals are able to digest as food, but no animals, including cows, can digest them when they are connected together, as they are in the form of cellulose or hemicellulose. Animals cannot produce the enzymes, notably cellulases, necessary to break down cellulose and hemicellulose into their component sugars. Many microbes, however, can produce cellulase. That is the basis of the strategy that cows use to live on a diet of grass: harbor microbes in their digestive tracts to break down cellulose for them into usable sugars. The microbes live on some of these breakdown products of cellulose; the cows live on other breakdown components and on the microbial cells as well.

Cattle, of course, are not the only animals that house microbes in order to live by grazing largely on cellulose. All ruminants, a group of about 150 species that includes sheep, goats, deer, elk, bison, buffalo, wildebeests, antelope, camels, and llamas, do as well. Another group of herbivores, called *cecal digesters,* includes horses, donkeys, rabbits, guinea pigs, and certain monkeys. Although they possess a different intestinal anatomy, they also use microbes as digestive aids, albeit somewhat less effectively. Ruminants have evolved elaborate modifications of their digestive tract in order to accommodate the enormous quantity of microbes needed to digest their food. They have a four-compartment stomach.

The first compartment, the rumen, is where the microbes are housed. This huge culture vessel in cattle holds about twenty-two gallons of liquid. Even the much smaller sheep has an over-six-gallon rumen. The numbers of microbes in the rumen are staggeringly large. Although quantities vary depending on the animal's diet (in general they harbor larger numbers when fed on forage rather than on grain), an average estimate seems to be a bit more than 10 billion per gram of contents. So in round numbers, a twenty-two-gallon cow's rumen would be home to about a quadrillion microbes, about 200,000 times as many microbes as there are humans on Earth, or several times our national debt in pennies. As I mentioned, the principal and critical role these microbes play is breaking down cellulose into its component sugars (largely glucose). This process is a major metabolic challenge, largely because cellulose is so extremely insoluble in water. Starch and glycogen, which animals, including us, can readily use as nutrients, are also macromolecules composed of long strings of glucose molecules linked together in a different way. But we, other animals, plants, and microbes are all able to produce the enzymes (amylases) to break the bonds of these more highly soluble polysaccharides into their component sugars.

Even the small bits of cellulose that the cow makes by chewing its cud are much too large to be taken into a microbial cell for digestion. So to attack such solid chunks of cellulose, microbes must release the enzyme, cellulase, outside their cells so it can attach directly to the surface of a particle of cellulose—the smaller the particles, the greater the total area of surface that is exposed to microbial attack. That is why the cow has to grind cellulose in its feed into small pieces by chewing its cud: not for microbes to be able to consume it as such but to create a larger surface area for the microbe's cellulase to attach to and thus digest fast enough to supply the cow's nutritional needs.

The other three compartments of the ruminant's stomachs also play essential roles in the cow's use of cellulose. The second smaller compartment, the reticulum, with about a four-gallon capacity in the case of cattle, is really an extension of the rumen, a mixing chamber. It also adds to the rumen's effective microbe-cultivating volume.

Together these two stomach vessels constitute a churning fermenter in which the cellulase that the microbes produce cleaves cellulose into molecules of soluble sugars, principally glucose. The cow never gets a chance to use these sugars, however. They are greedily taken up and utilized by the rumen's microbes for their own benefit. But because the rumen is anaerobic and lacks adequate supplies of oxygen substitutes to use for anaerobic respiration, these microbes can only ferment the cellulose-derived sugars. (Any oxygen that enters the rumen along with ingested food is rapidly depleted by the dense population of microbes that live there.) The fermenting microbes produce largely organic acids as endproducts, such as acetic, propionic, and butyric acids, which the cow absorbs through the walls of its rumen and uses as nutrients. The microbes cannot compete with the cow for these organic acids because they are unable to use them under the anaerobic conditions that prevail in the rumen.

The next of the ruminant's four digestive compartments, the omasum, is lined with folds. By means of muscular action, it creates a vacuum that pulls finer components of the rumen's contents out through a small orifice, leaving the larger particles behind to be digested further. In addition, the omasum absorbs water out of the rumen mix before it passes into the fourth chamber, the abomasum.

Here we are on more familiar ground. The abomasum functions somewhat like our own stomach. It is an acidic, enzyme-filled organ capable of digesting the microorganisms produced from cellulose in the rumen. These microbes are the ruminant's major source of protein. Thus the cow consumes grass, but it lives on the products of microbial fermentation: organic acids and microbial cells. A cow and other ruminants are in essence mobile factories that house a fermentation tank that they constantly supply with ground-up cellulose, using its products as nutrients. All classes of microbes—bacteria, archaea, protozoa, and even fungi—are present in the rumen and contribute, each in their own way, to the ruminant's and their own co-evolved lifestyle, forming a self-contained, mobile food chain.

By attacking cellulose, bacteria are the principal first links of the chain. Methanogenic archaea are the second. They add to the microbial biomass of the rumen and thereby to the cow's nutrition when they are digested in the abomasum. These archaea eke out a

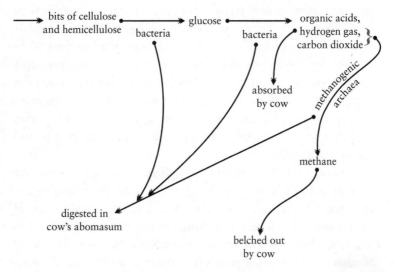

The rumen's core food chain.

living by utilizing hydrogen gas and carbon dioxide that other microbes discard as endproducts. Protozoa for the most part are further down the food chain; they consume bacteria, archaea, and occasionally one another. But, of course, eventually the cow consumes them. Some protozoa, however, are further up the chain because they do produce cellulase. Fungi, although few in number, make their own special contribution by attacking, albeit slowly, lignin, the component of plant material that is the most resistant to microbial attack. The fungi in the anaerobic rumen are necessarily an unusual breed. Although the vast majority of fungi are obligate aerobes, i.e., they cannot reproduce without oxygen, those living in the rumen are anaerobes out of necessity.

The microbe soup consisting of all major classes of microbes living together in the anaerobic rumen produces a highly distinctive, never-to-be-forgotten smell. To most humans (with the exception of a few dedicated rumen microbiologists) it is a stinking mix of offensive odors—but not so to canines. Canine predators tear open and consume the rumen contents of their herbivore prey before they start on its flesh. Perhaps evolution has taught them what a rich source of vitamins it is. The methanogenic archaea, for example, are abundant producers of vitamin B12. Manufacturers of dry dog food are well aware of this unusual canine food preference. Dry dog food, a highly processed substance unlike anything a dog evolved to consume, is not attractive to dogs on its own, so it is treated in a variety of ways to confer a rumenlike odor, though much less intense than the real stuff. The lower intensity is probably a compromise between canine preferences and owners' tolerances. If you are curious about the odor of rumen fluid, smell some dry dog food and imagine its being much more intense. Some manufacturers of dry dog food have even experimented with cultivating rumen bacteria and adding the culture mix to their product.

In spite of its odor, some humans have benefited from rumen contents as a dietary supplement. Certain tribes of Eskimos who live almost exclusively on reindeer, the flesh of which alone would constitute a vitamin-deficient diet, do not suffer such deficiencies as scurvy, which affected early explorers of the region, because they supplemented their diets with vitamin-rich rumen contents from reindeer.

I mentioned earlier that certain nonruminants, including such diverse groups of cecal digesters as equines, rodents, and simians, also benefit from the cellulose in their diets. Perhaps even we do, a bit. Like ruminants, cecal digesters depend on a population of cellulase-producing microbes in their intestinal tracts, but their intestinal anatomy is quite different from that of ruminants. Cecal digesters have no rumen in which to cultivate microbes. Instead they have a markedly enlarged cecum—that pouch connected to the intestinal tract between the small and large intestines—the structure to which our appendix is attached.

This cecum, constituting about 30 percent of the large intestine of a horse or about 18 percent of its entire intestinal tract, functions rather like a rumen: it is populated with similar microbes doing the same sorts of things. There are certain advantages to the cecal system over a ruminant one. For example, protein and vitamins can be used directly by the animal rather than being monopolized by microbes because they are absorbed in the small intestines before they enter the microbe-filled cecum. But there are distinct disadvantages as well. Because the cecum is located in the intestinal tract beyond the stomach, the microbes produced in the cecum cannot be digested. They pass into the feces unutilized. However, some rodents are markedly coprophagious (eat their own feces). Perhaps this behavior developed to recover some of the nutritional benefits of these microbes. Certainly the practice is beneficial for rodents: the growth of rats and rabbits in a natural environment is stunted significantly if this behavior is prevented. For most cecum digesters, however, only the products of microbial fermentation can be utilized as the contents of the cecum empty into the colon, where adsorption occurs but digestion does not. So most cecal digesters such as horses and donkeys are not able to live on grass alone. Their diet must be enriched with grain or some other, richer nutrient.

Housing microbes in the gut as a digestive aid for cellulose is not limited to mammals. Some insects coevolved similar mutualistic symbioses with microbes. Cockroaches benefit nutritionally from the cellulose they consume. But termites are clearly the champions. They live on woody plants, both living and dead, that are composed largely of a mixture of cellulose and lignin. And termites consume vast quantities of them. Some small plants in the American Southwest are completely consumed by termites while still standing erect. The collective damage termites cause to wooden structures is immense, an estimated 2 to 3 billion dollars annually in the United States alone.

Termites are a socially organized destructive force, living in huge colonies with reproductively capable queens and kings together with sterile soldiers and workers. The king termite assists the queen termite to create and attend to the colony during its initial formation, then continues to mate with her throughout his life. Colonies live in huge earthen or wooden nests or in buildings. One successful colony leads to more. The reproductive forms can fly to a new location, lose their wings, and settle down there to start a new community.

Like the burping cow, termites can be heard as well as seen. If you sit quietly in a house in some subtropical location, regions of Hawaii for example, you can clearly hear termites chewing away. I recall having dinner at a friend's house in the rain-soaked hills above Honolulu, pauses in the conversation filled with the crunching sound of termites chewing away at the wooden structure.

Termites and cows use the same basic nutritional strategies, but there are, of course, dramatic differences. The most obvious is scale. As we noted, the cow's microbe-packed fermentation vessel, the rumen, holds about twenty-two gallons. The termite's vessel, its swollen hindgut, holds approximately two-tenths of a milliliter—about four thousandfold less. And the microbial metabolism that occurs there is also different—different enough that the termite does not need to belch. For reasons that are not too clear, possibly because the termite's hindgut is a bit more acidic than the cow's rumen, most termites do not house significant numbers of archaea nor produce much methane, although primitive soil-eating termites do.

Like those in the cow, the cellulase-producing microbes in the termite initially attack bits of cellulose in the gut, producing hydrogen gas and carbon dioxide as endproducts. In cows, as we have noted, these are the starting material for methanogenic archaea, which then produce methane. In termites these products of the cellulose breakdown are used by acetogenic bacteria as starting material to make acetate (neutralized acetic acid—vinegar), an exclusively microbial, metabolic energy-acquiring scheme called acetogenesis.

Producing acetate instead of methane is a vital difference that fits well with the termite's lifestyle. Cows and other ruminant animals benefit from the methane formation, even though they must belch out the endproduct, because they can digest the archaeal cells that are produced as a result of the process. Termites cannot digest the microbial cells produced in their gut. They live almost exclusively on the acetate produced by the consortium of microbes they house.

A variety of animals, then, including all the ruminants, cecal digesters, and insects we have talked about in this sighting are richly rewarded by the unique ability of microbes to attack and degrade nature's most abundant organic product, cellulose. We humans are beginning to realize or, perhaps, are being forced to realize that we should not be left out. Our lifestyles and perhaps even our continued existence might depend on it. Unlike other organisms, we humans use and have become dependent on energy from sources other than our own metabolism. The vast majority of this energy is derived from the metabolism of other organisms, most of it having occurred in the ancient past. The fossil fuels we consume are just that. They are the convenient, ready-to-use remnants of organisms, principally plants and algae, formed by ancient microbial and geochemical activities. We have acquired an unsustainable habit. Sooner or later we will have depleted nature's reservoir of these irreplaceable resources. But, even more critically, by using these resources we are creating massive imbalances in the distribution of carbon on our planet, moving much of it into the form of atmospheric carbon dioxide, with its well-known blanketing and warming consequences. Some call it catastrophic climate change. On their own, even microbes, which have so effectively kept this planet in life-sustaining balance, do not seem able to bail us out.

One possible solution, which is being actively pursued in research supported by the U.S. Department of Energy, is to solicit the help of microbes more directly—to use cellulose-degrading microbes. Those from the hind gut of termites that degrade cellulose into ethanol seem particularly attractive. They would help us create a balanced cycle of carbon: carbon dioxide used by plants would make cellulose, which would be converted by microbes into ethanol, which we could burn and produce carbon dioxide. Humans have long benefited from the cow's exploitation of microbes; perhaps we will also learn by example and benefit from the termite's slightly different mode of exploitation.

A Fat Man and a Thin One

The contrast of a fat person and a thin one intrigues us. Comedy (Laurel and Hardy), literature (Don Quixote and Sancho Panza), and nursery rhymes (Jack Sprat and wife) exploit it endlessly in lighthearted ways. In view of the current obesity and linked diabetes epidemics, however, such a pair raises more urgent questions about the root cause of their differences. Does one of them eat much more and exercise significantly less, or might the difference depend on something within them? Might the difference be a microbe sighting? Recent research tells us that it is, in part—the two have different populations of microbes in their intestines, and their microbes perhaps contribute to their contrasting weights. The research also suggests that we are more like the cow and other ruminants than we had previously suspected, that microbes in our intestines like those in the cow's rumen or the horse's cecum digest some substances in our diet that we otherwise could not, thereby augmenting our food's caloric value.

In our species' distant past, this added morsel of energy might well have tipped the delicate balance between starvation and survival. Today in a more food-abundant world, at least for many of us, it might tip the scales toward obesity and diabetes. In the West, more than 500 million people are overweight and 250 million of those obese. A whopping 64 percent of Americans are overweight, and nearly half of these are obese. Such figures are enough to cast the once life-saving microbes in a different light—as pariah microbes that rob us of our good health, appearance, and mobility. The price paid for obesity in the United States is high. The U.S. Department of Health and Human Services estimates it may be responsible for 300,000 deaths annually, qualifying as number two, just behind cigarette smoking, in the ranking of preventable causes of death. Should we try to rid ourselves of these microbes? The root cause of obesity, of course, is our consumption of more calories than we expend, but microbes might add a few more calories to the food we eat than those listed on the label.

Humans are a mixture of our own cells and microbial cells living together. In certain respects, microbes constitute the larger portion. The total number of microbes on us and in us is ten times greater than all our own human cells. Ten to a hundred trillion microbes live within our intestines alone. And the total quantity of metabolic information encoded in the genes of our microbial guests is many times greater than in our own genes. Their metabolic potential exceeds ours.

Microbiologists have long speculated about the harm or benefit that our vast population of intestinal microbes might confer. Elie

Metchnikoff, the great Nobel Prize–winning microbiologist of the late nineteenth and early twentieth centuries, was convinced that our intestinal microbes do us no good, that our large intestine is "an asylum of harmful microbes, . . . a reservoir of waste of the digestive process and this waste stagnates long enough to putrefy. The products of putrefaction are harmful." Indeed, he believed that the large intestine is a physiologically unnecessary organ, merely a holding chamber uniquely well developed only in mammals because they alone "lead an extremely active terrestrial life," so the "need to stop to empty the intestines would be a serious disadvantage." Metchnikoff advocated consuming large quantities of yogurt, as the long-lived Bulgars did, to rid the large intestines of "products of putrefaction." Metchnikoff's influence lingers to the present. Yogurt as well as related lactic-acid fermented milk products are still touted for health and longevity, as is "colonic cleansing." Little scientific evidence supports these convictions.

There is, however, ample confirmation that the lactic acid–producing bacterium, *Bifidobacterium*, which predominates in the intestines of breastfed infants, does protect them from diarrhea-causing bacteria and their deadly consequences.

Arguments about the possible harm or benefits conferred by our intestinal population of microbes began to fade when it became possible to raise "germ-free" (axenic) mammals. For example, mature mouse fetuses could be taken aseptically from their mothers, fed sterilized food, and kept in isolation chambers. These germ-free animals appeared quite normal in all respects, with good health and normal life spans.

The perceived connection between weight gain and intestinal microbes rests on some experiments by a microbiologist, Jeff Gordon, that are difficult to dispute. He introduced the mixture of microbes from the ceca of normally raised mice into germ-free mice. Without changing their unrestricted access to food, the "conventionalized" mice gained weight while eating somewhat less food. Further, he compared the weight-producing power of intestinal microbe pop-

ulations of obese mice with those of normal-weight mice. Again
the results were convincingly clear. The population of microbes of
obese mice caused greater weight gain than the population from
their normal-weight littermates. Obesity of the mice used in these
experiments was genetically determined. The obese mice lacked the
ability to make leptin, the hormone that signals satiation. As a consequence they would overeat to obesity.

With a few assumptions and analogies drawn from the cow, these
results do make solid microbiological sense. We consume a number
of carbohydrates (polysaccharides) that we are genetically unable
to split into their utilizable sugar components. But microbes (principally bacteria) in our large intestine can split them, and in that
anaerobic environment, ferment their sugar components, producing endproducts (small organic acids and alcohols) that we are able
to take up and utilize as nutrients. Thus microbes in our intestines
could convert some dietary nutrients that we cannot use into other
nutrients that we can. Detailed comparisons of the microbial populations in the intestines of fat and thin people support this concept.
Taking a comprehensive microbial census of such a population
by conventional microbiological culture methods would constitute
a overwhelmingly immense enterprise, but modern approaches
(sometimes called *metagenetics*) of sequencing microbial DNA in
the intestines gives a clear picture of the classes of microbes that are
present. Such studies reveal that fat people have a greater proportion of certain kinds of bacteria (those belonging to the division
Firmicutes) than thin people do. And members of this division have
enhanced capacity to split and ferment polysaccharides that we
alone cannot as compared to the bacteria (belonging to the division
Bacteroidetes) that they replace. That all makes excellent sense. Fat
people get more calories than slim people from the same diet because their intestinal microbes have a greater capacity to utilize certain polysaccharides. But an important caveat emerges from these
studies. If a fat person loses weight by dietary restriction of any

sort, his or her population of intestinal microbes shifts. The shift in microbial population seems to follow rather than cause change in weight. Or to put it in other terms, the observed shifts in intestinal microbe populations do not fit a logical pattern of an evolutionarily selected adaptive change to protect our predecessor from starvation.

Methane-producing archaea also play an intriguing role in the fat-person-to-thin-person shift of intestinal microbe populations. Years ago a gastroenterologist, Michael D. Levitt, who is sometimes called "The King of Flatus," developed a quick and reliable way to determine if an individual produces significant amounts of methane in addition to hydrogen and carbon dioxide as an intestinal gas. He found that only about a third of humans produce methane. Because his method depends on measuring methane in exhaled breath, which in turn depends on intestinal methane being transferred to the blood and then in the lungs to the exhaled breath, small amounts of intestinal methane are not detected by Levitt's test, however. Later, in a detailed series of studies, Meyer Wolin showed that the distinction between methane- and nonmethane-producing humans is quantitative, not absolute. We all have methane-producing archaea in our intestines, so we all produce at least some methane. But most of us house only very small numbers of them and score negative on the breath test.

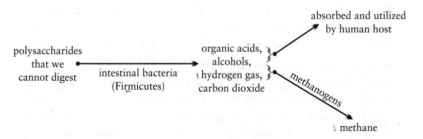

Food chain in the intestine of the fat man.

But does a person's population of methane-producing archaea contribute to yielding more calories from the food we eat? All these archaea do in the intestine is utilize two gases, hydrogen and carbon dioxide, while producing a third, methane; none of these gases can be used by us to generate calories. But the archaea could add more calories to our diet. They could do so by augmenting the amount of energy-yielding organic acids and alcohols that the fermenting bacteria make from the breakdown-product sugars from polysaccharides. By using one of the products of these fermentations, hydrogen, they enable the fermenters to make more of the others, namely, the organic acids and alcohols that we can use. Chemists call this phenomenon shifting the equilibrium of a reaction.

Because archaea are so fundamentally different from us and other microbes, it would be feasible to eliminate them from our intestines with drugs. Possibly, that would take calories from our diet and make the fat man more like the thin one.

A Gall on a Grapevine

A grapevine does not seem to be a probable location for a microbe sighting, but it most definitely is. If you look at the trunk and branches of any tree, shrub, or vine as you walk in the woods, a garden, or even a vineyard, you will notice that swellings on them are rather common. Insects cause some of these swellings, especially those that are almost spherical. They, too, are probably a microbe sighting by reason of having associated microbes that make the plant growth hormone indoleacetic acid, from the amino acid tryptophan, but that is another story. Other swellings are simply burls, which means that their cause is unknown, but often they seem to grow in response to some sort of injury. These plant growths are prized by woodworkers for their intricate and beautiful grain structure. Still other swellings have a peculiarly rough surface, and some of them are soft. These are crown galls and they are caused by a specific microbe, the bacterium *Agrobacterium tumefaciens.*

Crown gall has an important story to tell us about microbes and their intimate genetic relationship with higher organisms, in this case plants. We tend to think of evolution, as Darwin first conceived it to be, as an essentially linear process: Over time representatives of one species accumulate sufficient small, naturally selected genetic variations to become a new species. Then, the species begins to accumulate further variations. In modern terms, that implies that the genes we find in extant organisms are related to and derived from the genes of their lineal ancestors, although Darwin did not think in these terms. It must be remembered that he was not aware of Mendel's experiments with peas and therefore did not take into account the principles of modern genetics, including the concept of genes. Studies on crown gall revealed something new: microbes can donate some of their genes directly to modern organisms. So some genes in a plant, for example, might have been only recently acquired from a microbe, not necessarily derived step by step from microbes in dim evolutionary history. Crown gall might look like nothing more than a swelling on a plant, but it has huge implications about how evolution works. And it also is having a major impact on modern biotechnology.

Crown gall's cancerlike swellings are also known as plant tumors, probably because some of them continue to grow, becoming grotesquely large. But unlike animal tumors, they rarely kill their host. And for the most part, they do not even do significant damage. They can, however, weaken the stems of small plants, causing them to break. And because they are rather unsightly, they are particularly unwelcome on ornamental plants. So some considerable effort, with remarkable success, has been spent on ways to prevent crown gall—even to eliminate it from established trees. As we will see, crown gall can be effectively prevented in certain controlled environments, such as those in a nursery.

This particular plant growth is called crown gall because it frequently develops at or near the plant's crown, where stem and root join (and where ancients believed the plant's soul to reside—we are not so sure today), although crown galls can develop anywhere— on roots, and as in the case of our sighting, well up the trunk of the tree. Preference for the crown is a direct consequence of the fact that *Agrobacterium tumefaciens* is a normal inhabitant of the soil, particularly the soil around plants. Indeed, it has a distinct preference for the soil that surrounds a plant's roots, the region called the *rhizosphere*. *A. tumefaciens* enters the plant through minor wounds, which occur with greatest frequency near the crown as a result of trauma during planting, for example.

A. tumefaciens shows little preference among its menu of possible victims. Almost all broadleafed plants, as well as some conifers and even a few grassy plants, will do. This almost total lack of preference is startling in view of the extremely intimate interactions between the bacterium and its victim, the ultimate intimacy, perhaps: they mix their DNAs. The bacterium offers the plant a bit of its DNA, which the plant accepts and integrates into one of its chromosomes, making the donated bacterial genes a permanent part of the plant's own set of genetic instructions. Unsurprisingly, the plant's newly acquired genes continue to do the bacterium's bidding. They instruct the plant to create a hospitable habitat for *A. tumefaciens* to replicate, and to make special food for it, food so special that only *A. tumefaciens* is able to utilize it.

You will undoubtedly note that *A. tumefaciens*'s mode of manipulating its host is strikingly similar to a virus's modus operandi, working its will on its victim through genetic instructions that order the host to alter its activity to benefit the attacker.

A. tumefaciens carries its package of plant instructions in a very microbelike way—encoded in a plasmid, a circular bit of DNA. Many, probably most, bacteria carry plasmids. These small packages of genetic information play important roles in microbial biology, especially the biology of bacteria. We might ask what a plasmid is and how it differs from a chromosome, the familiar DNA package that all organisms, including bacteria, carry. We carry forty-six chromosomes; most bacteria carry only one. Most notably, a bacterial cell's plasmid is distinguished from its chromosome or chromosomes by being much smaller. Being circular is not a distinction. The chromosomes of most bacteria, including *A. tumefaciens,* are also circular. But *A. tumefaciens* is exceptional in that in addition to its circular chromosome, it also has a linear chromosome. The reason for this highly unusual situation is not understood. In general, however, only those bacteria that have linear chromosomes have linear plasmids. Either way, shape does not distinguish plasmid and chromosome.

The defining characteristic of a plasmid is its dispensability. Plasmids can be lost without doing great damage to the cell. In contrast, a bacterium cannot survive the loss of its chromosome because it carries vital instructions for metabolism, growth, and reproduction. Plasmids carry only highly specialized instructions for useful but not vital activities. If a bacterium loses a plasmid, its repertoire of activities and probably its competitiveness in nature are diminished, but the bacterium survives under most conditions. For example, many bacterial plasmids encode specialized instructions that make the bacterium resistant to antibiotics. If the bacterium loses such a plasmid, it becomes susceptible to the damaging or lethal onslaught of an antibiotic and cannot survive in an antibiotic-containing environment, but it does perfectly well in other environments. Similar disease-causing toxins, such as the anthrax toxin among many others, and pigmentation are usually plasmid-encoded as are a number of other skill sets of bacteria— for example, the abilities of species of *Rhizobium* to form mutualistic symbioses with plants, as we will see in the next chapter.

Many plasmids have yet another extraordinary ability, one that exerts a profound impact on microbial biology. They are able to move themselves from one cell to another. Some such self-transferring plasmids can transfer themselves only to other bacterial cells that are closely related to the one in which they reside. But others, termed *promiscuous plasmids,* have an almost unlimited capacity of self-transfer. Some, for example, can transfer themselves among all gram-negative bacteria. This remarkable, almost scary ability is one of the bases for the rapid spread of antibiotic resistance among bacteria. When such resistance arises and is encoded on a promiscuous plasmid, it can spread rapidly to any other gram-negative bacterium. For example, if the antibiotic resistance arises in bacteria in the intestines of a pig being fed antibiotics so it will grow faster, and the resistance is encoded on a plasmid, it can spread rapidly to other bacteria outside the pig, including those that cause diseases of humans, making that antibiotic less useful for human medicine. Unnecessary use of antibiotics for any purpose holds the threat of diminishing the utility of those antibiotics for all purposes.

The plasmid that tumor-causing strains of *A. tumefaciens* carry (termed "Ti" for "tumor-inducing") does even better. It can transfer some of its genes and integrate them into organisms belonging to another kingdom and domain, namely plants.

But first *A. tumefaciens* must enter the plant. When some of its cells come into close proximity of a wound on a plant, an elaborate interplay begins between these two distantly related organisms.

Chemicals that the plant releases after having been wounded, namely, phenols (aromatic alcohols), certain sugars, and acids, signal the beginning of this intimate plant-microbe relationship. *A. tumefaciens* cells are powerfully attracted to the wound by these chemicals, reminiscent, perhaps, of a shark sensing blood in the water and lunging toward its source. *A. tumefaciens* can detect and respond to one of the phenols (acetosyringone) at concentrations as low as 0.00002 milligrams per milliliter. It does not take much of this blood in the microbe's waters. These chemicals also activate certain genes (called *vir,* for "virulence") on the bacterium's Ti plasmid, genes that are quiescent when the bacterium lives alone. When activated, *vir* genes start to make a set of proteins (Vir proteins) that bring about the transfer of some of the genes on the Ti plasmids to the plant. But first the bacterial cell attaches itself firmly to the wall of a plant cell and builds a remarkable and highly specific, gene-carrying bridge between the two cells—bacterial and plant.

Next, another Vir protein clips a linear piece (the T-DNA) from the circular Ti plasmid, not just any piece but the specific piece bounded by runs of DNA called *border sequences,* and gene transfer is under way. The border sequences determine where clips occur. The DNA between them does not matter. As we will see later on, this seemingly trivial fact is vital to the biotechnological uses of *A. tumefaciens.* Other Vir proteins coat the T-DNA to protect it from DNA-destroying enzymes (DNases) that all cells contain, and escort it through the bridge into the plant cell. There it enters the cell's nucleus and is integrated at random somewhere into one of the plant cell's chromosomes. Then this former microbe gene becomes part of the plant's genome.

In a sense this is a gene invasion, a genetic coup d'état. The former microbial genes subvert the plant into housing and nourish-

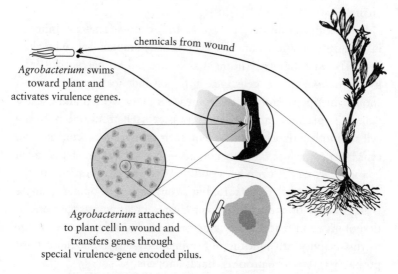

Agrobacterium transfers some of its genes to a plant.

ing the cells of *Agrobacterium tumefaciens* that are lurking in the wound. They direct the plant cell to make plant growth–stimulating hormone and as a consequence proliferate, creating the gall, which serves as a spacious dwelling for *A. tumefaciens*. Then comes the issue of nourishment for the invading bacterial cells. The plant cell's newly acquired genes instruct it to start making two sets of compounds: unusual amino acids (opines) and sugars (agrocinopines), a distinctly Machiavellian trick. Only *A. tumefaciens* and a few other microbes can use these compounds as nutrients. The plant cells cannot benefit from their own efforts.

Opines play yet another curious role in the complex relationship between *A. tumefaciens* and its plant host. The excess opines, unused as nutrients by *A. tumefaciens* cells in the developing gall, enter the plant's vascular system and some of the mixture is excreted into the soil surrounding the plant's root, i.e., into the plant's rhizosphere. The nearby, free-living *A. tumefaciens* cells benefit in a straightforward way from being supplied with nutrients that they and only a few other microbes can use. They prosper and proliferate in the soil.

But the impact on the Ti plasmid and *A. tumefaciens*'s ability to invade other plant wounds is even more complex and wily than that. The vast majority (in most soils, around 90 percent) of free-living *A. tumefaciens* cells lack a Ti plasmid, so they are unable, should the opportunity arise, to invade a plant wound and cause a gall. A Ti plasmid can, however, transfer itself from one bacterial cell to another that lacks one. For such transfer to occur, the two cells must come into intimate contact—thus such a transfer is called *conjugative*. Under ordinary conditions conjugative transfer of the Ti plasmid appears to be rare, but being bathed in opines changes that. An orgy of conjugal transfer ensues, creating a large population of invasion-ready *A. tumefaciens* cells. The selective advantage of this activity of opines is not abundantly clear. But there most probably is one. The proof is that it happens.

Although crown gall does not kill and rarely even seriously damages its plant host, it can be extremely unsightly. And crown gall can become epidemic in a nursery. The conditions there are just right. Plant propagation inevitably inflicts small wounds. Grafting, which is common in a nursery, causes a major wound. Also, as we have seen, galls markedly increase the populations of virulent *A. tumefaciens*. And for some plants, notably grapevines, infection-ready *A. tumefaciens* are shipped worldwide along with grape scions (graftable cuttings). That this dispersion actually occurs is supported by the fact that an identical strain of *Agrobacterium tumefaciens* (biovar 3) is found in crown galls on grapevines everywhere in the world.

Fortunately, however, an effective, inexpensive, and environmentally friendly method is now available for nurseries to prevent crown gall. The treatment is preventative, not curative, but it is highly effective. At planting, nursery stock or seeds are simply dipped in a container of water that contains a suspension of cells of a strain of *A. tumefaciens* that lacks a Ti plasmid. Such strains of *A. tumefaciens* are frequently given a different species name, *Agrobacterium radiobacter*, although the different species name does not seem to be justified because plasmids, including Ti, can readily be added to it by conjugation. And the plasmid can be removed from it by cultivation at an elevated temperature. In a sense, this mode of protection is a form of probiotic treatment. The particular Ti plasmid–free strain K84 (sold among other trade names as Galltrol) has the added advantage of producing a compound (agrocin 84) that is toxic to most Ti plasmid–bearing strains and to not much else. Such bacterial magic bullets that are toxic only to closely related bacterial strains are called *bacteriocins*. Many bacteria produce them, a reflection of nature's brutal competitiveness, one would assume.

Biotechnologists were quick to appreciate the remarkable potential of *A. tumefaciens* and its Ti plasmid for use as a tool in plant breeding: it has the capacity to add foreign genes to a plant's genome—not just the bacterial genes in the T region of the Ti plasmid, but genes from any source. It is only necessary using standard recombinant DNA technology to insert the foreign gene into the T-DNA region of the Ti plasmid. Then *A. tumefaciens* and the Ti plasmid will do the rest. But, of course, the plant breeder wants to breed a new plant, not just a new gall. To do this, the breeder must pass the genes into cells that will develop into a plant, not a gall. Cultured plant cells, called *protoplasts,* do the trick because protoplasts derived from most plants can be induced to develop into new plants. This approach, although for some reason controversial, has been used to develop a number of transgenic plants that are used worldwide.

As we have seen, *A. tumefaciens* can cause crown galls on a large variety of plants. It can transfer genes to cell cultures of a seemingly unlimited variety of kinds of organisms, including those from all kinds of plants, fungi, and even humans. Already new and useful varieties of corn, soybeans, cotton, canola, potatoes, and tomatoes have been developed using *A. tumefaciens* as a mechanism of gene transfer.

The gall on a grapevine is a microbe sighting with many ramifications; it is also an example of a particularly intimate symbiosis at the genetic level between a microbe and a plant.

Small Points of Light on a Dead Fish

This sighting requires some unusual preparation, but it is worth it. In a darkened room, gaze for awhile at an ocean fish, preferably one that has not been frozen, but has been in the refrigerator for several days. At first you will not see anything, but as your eyes adapt to the darkness, bright spots of light will begin to appear on the surface of the fish.

These are colonies of light-emitting microbes, most probably bacteria. Like other small piles of cells that microbiologists call "colonies," they are the progeny of a few or possibly a single microbial cell. We cannot see these bacterial colonies on fish as points of light until enough cells are present to attain sufficient size. One reason is obvious: the light source must be intense enough to perceive. But even if we were able to detect extremely small amounts of light, we would notice that one or a few cells do not light up. A group of them is needed to turn one another on. This microbial custom of cells in groups stimulating one another to do things that fewer cells cannot is called *quorum sensing*. As the name implies, certain bacteria have the ability to evaluate whether their population has attained a certain critical size and, if so, to do things that benefit the crowded population but would be a waste of effort for a sparser one. We will put off, for the moment, speculating about why it is advantageous for bacteria to delay lighting up until they sense that they are crowded. Microbes sense quorums for a lot of reasons.

Were we to isolate and characterize the bacterium that comprises these lit-up colonies, we would most probably discover that it was *Alivibrio fischeri,* a bacterium not too distantly related to the old familiar *Escherichia coli. A. fischeri* is also found abundantly, floating freely as individual cells (planktonic) in the ocean. This seems odd in view of their becoming luminescent only in crowds. Perhaps being planktonic is only a transitory phase of their lifestyle; their true home is in association with animals, such as fish or squid, where they accumulate in specialized structures termed *light organs,* areas of the marine animal that luminesce. If that is the case, light organs are their true home, and they are planktonic in the ocean only because they have been shed from such an organ.

Luminescent bacteria living in an animal's light organ participate in a remarkable mutualistic symbiosis: the animal supplies nutrients, and the dense population of bacterial cells supplies light. Various animals use such microbe-generated light for many different purposes, but not as we might expect, as a flashlight to guide their way in the dark ocean abyss. The manifold uses, including camouflage, attraction, avoidance, repulsion, and communication, are

startlingly diverse, some even seemingly imaginative and innovative.

The anglerfish is a particularly remarkable innovator. This deni-
zen of deep-sea darkness has a light organ at the tip of an appendage that extends over its head. Other small animals are fatally attracted by it to within the fish's striking distance. A shark plays a
similar trick. It has a light organ on its lower surface that it uses to
attract prey, themselves predators. These unfortunate smaller predators mistakenly interpret the small, lit organ as their manageable
prey, only to realize that they have become the shark's prey.

The small Hawaiian squid, *Euprymna scolopes*, only about an
inch and a half long, uses its light organ for quite the opposite reason—for protection, not aggression. This highly vulnerable little
creature hides during the day by burying itself into the sand flats
dispersed along its coral-reef home. But it must come out and expose itself at night in order to forage. Then it depends for protection on its light organ, which is located within its mantle cavity.
The squid precisely regulates the amount of light the organ emits
to render itself invisible from below, a process called *counter-
illumination*. It emits just enough light to balance the light from the
sky, enough to prevent casting a shadow but not enough to appear
bright. Either extreme would attract predators. The squid has to do
the adjusting itself because it cannot modulate the amount of light
that its bacterial guests produce in the light organ. The squid has
two light-adjusting tools, a reflective area on its lower surface and
an ink sack. To increase emission, it moves its light organ over the
reflective surface. To diminish emission, it tucks part of the organ
behind its opaque ink sack. Remarkably, the squid senses when the
intensity of counter-illumination is ideal.

The light organ and its proper adjustment is a complex as well as
vital part of this relatively simple animal's lifestyle. When the squid
is first hatched, its light organ contains no bacteria. But planktonic
Alivibrio fischeri cells, and only they, are soon attracted to it by

some still-unknown mechanism. They enter the organ through one of several entry pores and attach themselves to the surface of the organ's multifolded surface. The squid feeds them well, so they multiply rapidly, doubling their population about every thirty minutes until around 100 million bacterial cells populate the organ. Then *A. fischeri*'s quorum-sensing mechanism signals them to light up. The bacterial cells continue to multiply, but now quite slowly, only rapidly enough to replace the cells that are continually being sloughed, again to become planktonic.

The flashlight fish, *Photobletheron,* a native of the eastern Mediterranean Sea, also uses its microbe-filled light organ for protection, but in quite a different way. It has two coverable light organs on opposite sides of its head. One would think that they would attract predators and be a hazard, but because of the way that *Photobletheron* turns them on and off, they are not. *Photobletheron* zigzags constantly as it swims. Just before a turn, it covers its light organs, turning the light off. The lurking predator strikes where *Photobletheron* would have been, had it not turned. *Photobletheron* swims safely away, lights blazing.

So we see that luminescent bacteria in the ocean have evolved not to waste their metabolic energy during their solitary planktonic phase of existence, expending it only when being rewarded by the nutrients supplied by an animal's light organ. But there is a legendary, somewhat romantic exception.

Mariners have from time to time, often unconvincingly to listeners, described the phenomenon of the whole sea around them lighting up, a phenomenon that came to be known as *milk sea*.

Two crew members in Jules Verne's classic *Twenty Thousand Leagues Under the Sea* talk about it:

> The 27th of January, at the entrance of the vast Bay of Bengal . . . , about seven o'clock in the evening, the Nautilus . . . was sailing in a sea of milk . . . Was it the effect of the lunar rays? No: for

the moon . . . was still lying hidden under the horizon . . . The
whole sky, though lit by the sidereal rays, seemed black by contrast with the whiteness of the waters.

"It is called a milk sea", I explained . . .

"But sir . . . can you tell me what causes such an effect? For I
suppose the water is not really turned into milk."

"No, my boy: and the whiteness which surprises you is caused
only by the presence of myriads of infusoria, a sort of lumi-
nous will worm, gelatinous and without color, of the thickness of
a hair whose length is not more than seven-thousandths of an
inch. These insects adhere to one another sometimes for several
leagues."

". . . and you need not try to compute the number of these infusoria. You will not be able, for . . . ships have floated on these
milk seas for forty miles."

The milk sea remained in or close to the province of legend until
relatively recently. Ocean samples were taken after creditable sightings were made. But the ultimate certification occurred in 2005
with the publication in the *Proceedings of the National Academy
of Sciences* of an account of scientists being able to match the time
and place of the SS *Lima*'s sighting of a milk sea with bright spots
on satellite images from space. The images confirmed that off the
coast of Somalia a milk-sea bioluminescent event had indeed occurred on three consecutive nights. It covered an area roughly the
size of Connecticut.

And the samples established that luminescent bacteria were the
cause of the sea's lighting up. But how did they attain their required
light-emitting quorum? The primary cause is most likely a massive
bloom of algae, probably of the microalga *Phaeocystis*. Then the
luminescent bacteria reach a quorum either by forming colonies on
clumps of algae or growing to a dense enough population at the
expense of the nutrients that the algal bloom supplies.

But what about the luminescence that many of us have seen—the
lighting up of a ship's wake, a dolphin's bow wave, or a breaking

surf, especially in tropical waters? I experienced a particularly spectacular example of this phenomenon during World War II at quite a different venue—the crew's head [bathroom] aboard the USS *Cowpens*, CVL 25, then in the South Pacific. As in most naval vessels of that era, the crew's head, several decks down, was equipped as toilets with long metal troughs crossed with rows of wooden slat seats and constantly flushed with seawater scooped up by the ship's movement. When I turned off the lights to make what I thought might be a microbe sighting, I was rewarded, along with the howls of some very angry shipmates, with the spectacular sight of the toilet troughs glowing like huge neon tubes in the pitch black compartment—a particularly gorgeous microbe sighting.

These ocean waters luminesce thanks to protists called dinoflagellates, not bacteria. Certain species of dinoflagellates are intensely luminescent. Dinoflagellates are curious unicellular, eukaryotic mi-

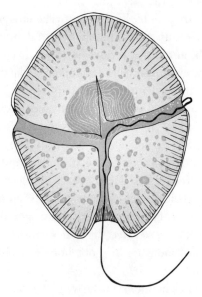

A dinoflagellate.

crobes (protists) that swim with a distinctive spiral motion propelled by two flagella. Most acquire their metabolic energy by
oxygen-yielding photosynthesis. They glow more intensely during
a night following a bright, sunny day that increases their energy
charge. But what about quorum sensing and why do they seem to
light up only in turbulent water? Quorum sensing is easy. They do
not practice it. Even a single cell, all alone, can light up. The re-
quirement for turbulence is good deal more complicated. It is not
that turbulence is required to aerate the water, which, of course, it
does. It is the shear force created by turbulence that flips the dinoflagellate's light switch. The precise mechanism remains a mystery,
but events leading to it are well established. Shear forces mechanically distort the cell, rearranging subcellular organelles in a manner
that triggers them to emit light. Perhaps, as some biologists speculate, this is a means of self-protection. The turbulence caused by an
approaching predator triggers the dinoflagellate suddenly to light
up, perhaps repelling the predator.

Certain fungi are also bioluminescent. The most notable and eerie, foxfire, may result in the illumination of damp logs in the forest, usually in the autumn. It is not easy to see this feeble, eerie, pale
green light. Your eyes must be fully adapted to the dark, and even a
bright moon or urban light pollution will render foxfire invisible.
But it has been observed and appreciated since ancient times. Aris-
totle noted "cold fire" in the woods, as did the Roman naturalist
Pliny on rotting olive trunks. Some have put it to practical use, to
find their way through a forest at night, for example. Mark Twain's
characters Tom Sawyer and Huckleberry Finn used a glowing log
as a light to dig their tunnel. A variety of basidiomycete fungi are
bioluminescent, most belonging to the white-spored group of gilled
mushrooms. *Armillaria mellea,* the honey mushroom, is the most
notable. The mushroom itself is not luminescent, but its rapidly
growing mycelium is.

Microbes did not maintain exclusive rights to bioluminescence,
although they undoubtedly invented it. Other creatures as well light

up on their own without depending on symbiotic microbes. The list includes fireflies, a variety of worms, centipedes, and millipedes.

Bioluminescence in fish and turbulent water may be widespread, but it remains a somewhat mysterious aspect of the life of microbes. In two respects—quorum sensing and interspecies symbiosis—it reflects the changes caused by living together.

An Aphid Feasting on a Tender Rose Leaf

Insects evolved on protein-rich diets. In order to survive, let alone prosper, they must be supplied from their diet with ten amino acids, the same essential amino acids, as they are called by nutritionists, that we need. But sufficient amounts of these amino acids are not present in the food that aphids eat. Aphids and other sap-sucking insects exist on a diet exclusively of plant sap, which is essentially pure sugar water with only small amounts of nitrogen (largely in the form of the single amino acid glutamine) and almost completely devoid of the ten essential amino acids. As any avid gardener, par-

An aphid eating and giving birth.

ticularly a rose fancier, well knows, aphids do indeed prosper none-
theless. Each spring, as soon as tender new rose leaves appear, they
quickly become covered with sap-sucking aphids that spend their
lives there. How do aphids survive on such a diet? A microbe, a
symbiotic one, is the answer. A bacterium living inside the aphid's
cells supplies all the essential amino acids that the aphids cannot
live without.

Aphids are sap-consuming machines. When they settle on a leaf,
they push the needlelike tip (stylet) of their snoutlike feeding organ
(proboscis) between and through cells directly into the plant's sap-
carrying vessels (phloem). Because the vessels' contents are under
pressure, the aphid doesn't even have to suck. Sap just flows into it.
The process may be passive, but it is very effective. It has to be because the aphid must process vast quantities of sap to acquire adequate amounts of nitrogen. (One wonders why aphids did not capture a nitrogen-fixing bacterium as well as an amino acid–making
one.) Aphids separate out the nitrogen and secrete the excess sugar
as a sticky liquid—honeydew—which, of course, doesn't go to
waste. The honeydew spreads over the surface of the infested leaf,
and some drips onto lower leaves. Frequently, a black yeast grows
on the sugar the honeydew contains, thereby conferring the leaf
darkening that is characteristic of many aphid infestations. Ants
are also fond of honeydew. So much so that some ants farm aphids
to be assured of a good supply of their sought-after food. Aphid
farming is a full-time job. In the spring, farmer ants carry newly
hatched aphids to favorable locations on a susceptible plant for
aphids to develop. (Most species of aphids are restricted to particular plants.) The ants protect their aphids by killing the larvae of la-
dybugs and other aphid predators. And like practical farmers, the
ants slaughter excess (from the ant's point of view) aphids to feed
to their own developing larvae.

Aphids are astoundingly effective aphid-producing factories. A
single aphid can produce many thousand aphid progeny during a
single growing season. Female aphids, or *stem mothers,* emerge

in the spring from over-wintering eggs (eggs only produce female aphids). They are immediately able to produce more aphids. They are not even delayed by having to mate because they already have female embryos within them. And these embryos also contain female embryos. Thus, the stem mother is already a grandmother as she emerges in the spring. When her embryos develop, they are born alive (viviparous). And they, too, proceed to produce more female aphids by parthenogenesis (no male is needed so far). Soon the stem mother's progeny is a large mass of aphids, sufficient to cover one or several leaves on our rose bush. Only late in the season are some males produced. They mate with females, which causes the females to switch from giving live birth to producing eggs that can over-winter and give rise to next year's crop of stem mothers and infestation of aphids.

The ready switching from producing females to producing males can occur because only a single kind of chromosome (X) determines an aphid's gender, in contrast to the two kinds (X and Y chromosomes) that determine ours. A female aphid has two Xs; a male has only one. So the late-season switch to producing males occurs simply because one of the Xs is dropped during parthenogenetic development.

Aphids' remarkable fecundity depends on a reliable and adequate supply of the ten essential amino acids, and that depends on an intracellular symbiotic microbe that all aphids (and other sap-sucking insects) have. It has long been known that certain specialized cells, called *bacteriocytes,* within aphids contain intracellular microbes or at least structures that look like bacteria. The role of these presumed bacteria was unknown, and they could not be cultured in the laboratory. But they were almost certainly bacteria because adding a bacterium-killing antibiotic to the aphid's diet made them disappear. Although it may have cured the aphids' apparent bacterial infection, the aphids did not benefit; instead they sickened and eventually died.

The treated aphids could be maintained, however, by feeding

them an artificial diet that contained the ten essential amino acids, strongly suggesting that producing the aphid's required amino acids was the bacterium's essential role. When the bacterium's genome was sequenced, it became clear how this tiny intracellular symbiont is able to feed the aphid. It has evolved to become a prodigious producer of amino acids and to do little else. For example, one of the ten essential amino acids is tryptophan. The bacterium's DNA contains sixteen copies of the genes encoding tryptophan's biosynthesis, whereas only one set is sufficient for free-living organisms. The extra copies allow the bacterium to supply the aphid. Thus, bacterium and aphid form a curious nutritional, mutualistic symbiosis. The aphid supplies the bacterium with a home and feeds it with plant sap. The bacterium supplies the aphid's essential amino acids. The bacterium, called an *endosymbiont* because it lives inside the aphid's cells, is a vital part of the aphid, which is transmitted from one generation to the next through embryos or eggs.

The bacterium's DNA sequence also revealed its lineage. It is indeed a bacterium. The particular one found in one aphid, *Schizaphis graminum,* now called *Buchnera aphidicola,* is closely related to our old friend *Escherichia coli.* In the evolutionary processes of specializing to supply essential amino acids, *B. aphidicola* lost many functions that became unnecessary in its new protective environment. Its genome size plummeted from about 4 million base pairs to its present 650,000. In so doing, it lost its capacity for independent living. For example, it is no longer able to make any of the amino acids that the aphid can make for itself. Instead, *B. aphidicola* concentrates its synthetic abilities on making the ten amino acids that the aphid cannot make and therefore must have.

The intimate endosymbiotic association between *Buchnera* and aphids probably began 150 to 250 million years ago when an ancestor *Buchnera* infected an ancestor aphid, making the aphid's sap-consuming lifestyle possible. Since then, bacterium and insect have coevolved. Now there are over 4,000 species of aphids, all of which have an endosymbiont. Dramatic evidence of this interdependent coevolution comes from comparing the family trees of aphids and their endosymbiotic bacteria (constructed from similarities of their ribosomal RNA sequences). They are almost superimposable, proving beyond doubt that bacterium and aphid did, indeed, evolve together after the initial capture of an ancestral endosymbiont by an ancestral aphid. Neither aphid nor endosymbiont can live alone.

One of the 4,000 species of aphids is the viticulturalist's dreaded *Phylloxera*, the "plant louse," which attacks and kills the roots of wine grapes *(Vitis vinifera)*. In the early twentieth century, *Phylloxera* (which came from North America) wiped out wine grapes (which came from Europe) in Europe and those in North America that were growing on European roots. Now almost all wine grapes are grafted onto the roots of American grapes that are naturally resistant to *Phylloxera*.

Of course, as farmers and gardeners well know, aphids are not nature's only sap-sucking insects. Others include such unwelcome plant guests as whiteflies, mealy bugs, and psyllids (jumping plant lice). These insects' ancestors met their nutritional challenge the same way aphid ancestors did—by acquiring bacteria to live within some of their cells to supplement their otherwise inadequate diet. So, any sap-sucking insect is a microbe sighting.

Acquiring endosymbiotic bacteria to perform certain metabolic tasks is hardly a new strategy for eukaryotes. The earliest ancestors of all of them acquired a bacterium to take on the task of aerobic respiration for it and its progeny, which include us. That endosymbiont evolved to become a mitochondrion, the intracellular organelle that mediates all eukaryotic respiration. The earliest ancestors of all photosynthetic eukaryotes—plants, algae, and photosynthetic protists—acquired a bacterium, a cyanobacterium, that was capable of photosynthesis to do that food-producing service for them. When we include these highly evolved endosymbionts, any eukaryote, plant, animal, or protist becomes a microbe sighting.

As far as we know, using microbial endosymbionts to accomplish a metabolic task that eukaryotes are unable to do on their own is limited to aerobic respiration, oxygen-yielding photosynthesis, and synthesis of vital amino acids. One wonders if there are more metabolic tasks as yet undiscovered. Certainly, there are excellent candidates, including nitrogen fixation and various forms of anaerobic respiration. As I suggested, if aphids had acquired a nitrogen-fixing bacterium as an endosymbiont, they would not have to produce so much honeydew. Of course, that would be bad news for ants and plant-dwelling black yeasts. But the discovery of microbes' potential is far from complete.

5

Cycling Nitrogen

We have repeatedly discussed microbes' manifold roles in maintaining Earth's chemistry in life-sustaining balance. One of these is supplying the forms of nitrogen-containing nutrients that various members of Earth's highly complex mix of diverse living things need. Over the next series of sightings we will trace this chemical supply network and see that it is cyclic, beginning with and returning to Earth's vast reserve of nitrogen in the form of atmospheric nitrogen gas. This chemical network is called the *nitrogen cycle,* and for the most part, microbes run it. We will begin our examination of this massive and interconnected global flow of nitrogen humbly by examining the roots of sweet peas.

Visitors Penetrating the Roots of Sweet Peas

If you pull a sweet pea plant (or some other legume) out of the ground and carefully shake off adhering soil, you will most likely notice dozens of small (about an eighth of an inch in diameter) globular swellings on the roots. If you cut one open with a knife or squeeze it, and it is red inside, you have proof of a microbe sighting. The juice appears to be and is indeed blood red because it contains a form of hemoglobin that is closely related to the red pigment that is in our and other mammals' red blood cells. Sometimes it is hard to deny our kinship to microbes. Just about every sweet pea plant will have such swellings on its roots, except for those growing in soil that lacks appropriate bacteria. In fact, the majority of leguminous plants, from the sweet pea–sized ones to trees, have such swellings. Legumes can be recognized by their bonnet-shaped flowers. Texans call lupine, a legume, "bluebonnet." The swellings on legumes' roots are known as *root nodules* and they are densely packed with bacterial cells.

Leguminous plants are highly successful and abundant. There are about 13,000 described species. Common representatives, in addition to peas, include beans, soya beans, alfalfa, clover, vetch, peanuts, *Acacia,* and even kudzu. Most of them produce similar nodules on their roots, although they do vary in size and shape. On some species the nodules are spherical extensions and on others they are cylindrical.

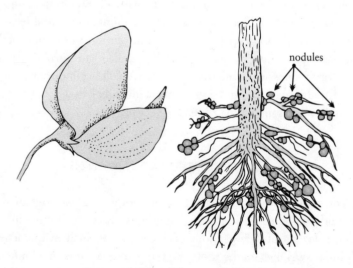

Root nodules and flower of a legume.

As dramatically unprepossessing as these structures are, their reach extends far beyond the legume that bears them. They exert a profound impact on the ecology of our planet because bacterial cells in them are the sites of nitrogen fixation, the conversion of nitrogen gas (N_2) into a fixed (nongaseous) form, ammonium ion (NH_4^+). Fixed nitrogen flows from these nodules through life's interconnecting chemical paths to all other forms of fixed nitrogen. All life forms require a source of fixed nitrogen because most of their essential components, including protein, membrane components, DNA, and RNA, contain it. All the nitrogen in all these various forms in all organisms was once nitrogen gas in the atmosphere.

Earth's supply of fixed nitrogen is not a stable resource that can be passed endlessly from one life form to another, such as happens when animals eat plants or microbes feast on the remains of dead creatures, because some fixed nitrogen is constantly being converted (by certain microbes, of course, and exclusively by them) into nitrogen gas and thereby being returned to the atmosphere. We will encounter these producers of nitrogen gas when I later discuss the products of a wastewater treatment plant. So, unless there is a continuous flow of nitrogen from its gaseous to a fixed form, all life would soon come to an end. Root nodules are not exclusively responsible, but they do account for a major portion of the natural component of the flow from gaseous to fixed nitrogen. Almost all the rest of the natural flow is mediated by free-living microbes and other microbes that form different symbioses with plants and even animals, for example, termites. The prokaryotic representatives of microbes, principally bacteria and a few archaea, are the only living things that are capable of fixing nitrogen.

Until the twentieth century, the natural activity of microbes supplied all fixed nitrogen on Earth with the exception of a relatively small flow mediated by spontaneous chemical reactions as a result of natural events, such as lightning strikes, ultraviolet radiation, and fires. Human innovations, such as internal combustion engines and other high-temperature processes, also contribute to fixing some nitrogen. But a major change occurred in 1908 when Fritz Haber, a German chemist, developed and patented a chemical process for converting nitrogen and hydrogen gases into fixed nitrogen in the form of ammonia (NH_3). During the First World War industrial production began. Now production of fixed nitrogen via the Haber process along with a few related industrial methods accounts for about half of the world's supply of it. Without question, the Haber process has been a major contributor to human welfare. The Green Revolution and its starvation-preventing consequences would not have been possible without the vast quantities of nitrogen fertilizer the Haber process supplies and their ability to increase crop yield. There is a downside, however. The process uses vast quantities of energy (principally in the form of natural gas), and the concomitant production of carbon dioxide as well as over-fertilization are ecological scourges. Witness the impact of the latter on the Chesapeake Bay, for example. Excess nitrogen and phosphorus in fertilizer, largely from farmlands, runs into the bay, stimulating prolific growth of algae ("blooms"), which deprive the underlying bay grasses of sunlight. The consequence is oxygen deprivation and death of marine life, including fish and plants. The bay is rapidly becoming an estuarial desert, to a considerable extent a victim of the Haber process.

Nodule-forming bacteria all belong to one of several closely related genera (*Rhizobium, Bradyrhizobium, Azorhizobium,* and *Sinorhizobium*). These rather ordinary-looking, rod-shaped bacteria live in the soil and can be readily cultured in the laboratory. They are extremely selective when it comes to choosing partners for symbiosis. Only particular species, indeed particular strains of particular species, form nodules on particular plants. And problems of compatibility extend beyond merely forming nodules. Some strains that form highly productive nodules on one legume will form significantly less productive ones on another. For these reasons, modern agricultural practice arranges optimal introductions between microbes and plants. Pure cultures of selected strains of rhizobia, cultivated commercially in large batches, are introduced to their appropriate plant partner as a coating on their seeds or as a liquid suspension added to the soil at the time of planting.

Bacterial cells and plant then initiate a highly elaborate, evolutionarily choreographed ritual dance of back and forth stimulation leading to nodule formation and nitrogen fixation. Neither plant nor *Rhizobium* alone can fix nitrogen. The dance begins by the root of a developing leguminous plant exuding a bacterial attractant (a flavonoid) into the soil. Nearby cells of a nodule-forming bacterium can sense minuscule concentrations of this chemical signal and they migrate toward its source, the beckoning root. When there, they attach to a root hair (a single-celled, nutrient-gathering extension from the root). The root hair responds to their arrival and the chemical signals they make to announce it by curling at its tip and then proceeding to make a welcoming pathway, called an *infection thread,* which guides the bacterial cells through the root hair to the underlying plant tissues. There they multiply and stimulate the plant to make their nodule home.

Within the nodule, the bacterial cells transform themselves com-

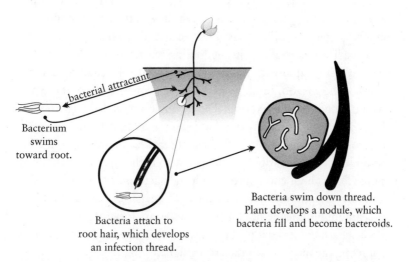

Bacterium
swims
toward root.

bacterial attractant

Bacteria attach to
root hair, which develops
an infection thread.

Bacteria swim down thread.
Plant develops a nodule, which
bacteria fill and become bacteroids.

Interactions leading to the formation of nodules and nitrogen fixation.

pletely, becoming nitrogen-fixing machines called *bacteroids*. Bacteroids differ in a number of ways, both morphologically and metabolically, from the bacterial cells that entered the root. They are thicker; they are branched; and they lack many normal bacterial functions, including the ability to divide and multiply. They become totally dedicated to their task of fixing nitrogen.

The plant also makes its vital contributions to their shared mission. It supplies carbohydrates that the bacteroids metabolize to generate the metabolic energy needed to drive the highly energy-intensive process of nitrogen fixation. And the process is indeed startlingly energy demanding. Twenty to thirty molecules of energy currency, ATP, are needed to reduce a single molecule of nitrogen gas to ammonia. Only a minor fraction of this energy is needed to drive the reaction itself. The majority of it is spent activating the highly inert nitrogen gas so it can react.

The plant also manages the delicately balanced issue of supplying oxygen to the finicky rhizobial bacteroids within the nodule. They demand just the right concentration of oxygen. Because rhizobia are obligate aerobes, meaning that they live by aerobic cellular respiration, they must have adequate oxygen to generate the considerable quantities of metabolic energy needed to fix nitrogen. But too much oxygen would be disastrous because nitrogenase, the enzyme that mediates the actual nitrogen-fixing reaction, is highly oxygen-sensitive: it is rapidly inactivated by only relatively small concentrations of the gas. So, too high or too low a concentration of oxygen would stop nitrogen fixation. The hemoglobin (called *leghemoglobin*) that the plant makes and that gives the nodule its characteristic red color accomplishes the challenging task of maintaining the intranodule concentration of oxygen within the narrow functional range that is high enough to permit the bacteroids to generate ATP and low enough not to destroy their nitrogenase. The plant adjusts the amount of leghemoglobin it makes to meet the bacteroids' needs. Farmers have long been aware that the redder a nodule's contents, the more fertile the crop will be.

Eventually, usually at the end of the growing season, the nodule deteriorates, spilling its contents into the soil. Bacteroids cannot survive without the plant's continued attention. They die with the nodule, but all nodules also contain a few dormant, rod-shaped, vegetative bacterial cells that did not differentiate to become bacteroids. When they are released back into the soil, they proliferate there, forming a population that is ready to respond, probably next year, to the beckoning call of another appropriate plant host.

Rhizobia's soil-to-plant-to-soil life cycle is well established and reliable, but it can benefit from human assistance. Naturally, some fields of leguminous crops do not become uniformly infected and, as we noted, the bacterium-plant symbiosis is so highly evolved that only a limited number of rhizobial strains are highly effective nitrogen fixers in a particular plant. That is why it has become common agricultural practice to introduce selected rhizobial strains to a particular crop either by coating seeds with them or adding cultures of them to soil during planting.

Managing leguminous plants to better crop yields is not exactly a recent human innovation. The Romans recognized the value of alternating crops of soil-depleting cereals with soil-enriching legumes, although the underlying scientific rationale for this wise practice, which continues in many places today, was not understood until the late nineteenth century. In 1886, two German agricultural chemists, Hermann Hellriegel and Hermann Wilfarth, proved unequivocally that growing legumes adds nitrogen to soil.

The rhizobia in the nodules on the roots of leguminous plants are important contributors to the world's supply of fixed nitrogen, but they are not the only microbes that form nitrogen-fixing partnerships with plants. Certain rhizobia belonging to the genus *Bradyrhizobium* form nitrogen-fixing symbioses with other plants, including the trees belonging to the genus *Parasponia* in humid tropical forests. Other bacteria besides rhizobia also fix nitrogen symbiotically with plants. One group belonging to the genus *Frankia* forms nitrogen-fixing symbioses with about two hundred different plants, including alder trees and *Ceanothus,* a genus of beautiful pale to bright blue bushy plants. The nodules *Frankia* forms on alders are huge and elaborate, as large as a human fist with many small lobes. Those it forms on *Ceanothus* are much smaller and harder to see than the ones on legumes. They are only about a quarter of an inch long and a third as wide.

Another bacterium, *Acetobacter diazotrophicus,* dwells in the stems of sugar cane, forming a productive association there that fixes around 150 pounds of nitrogen per acre. Certain cyanobacteria, a group of oxygen-producing phototrophs, as I will discuss shortly, also form nitrogen-fixing symbioses.

Still other bacteria fix nitrogen completely on their own in a free-living state. One of them, *Clostridium pastorianum,* is an anaerobe, so its nitrogenase is environmentally protected from oxygen's destructive impact. But another group of bacteria, *Azotobacter* spp., is composed of strict aerobes, which are dependent on a supply of oxygen to generate their metabolic energy by aerobic respiration. They protect their nitrogenase in a curious and unexpected way: by respiring oxygen at their cell surface so rapidly that the interior of the cell, where its nitrogenase is located, is kept free of oxygen. And certain archaea living in the oxygen-free termite's intestine satisfy a bit of this animal's need for fixed nitrogen.

In spite of industrial nitrogen fixation's massive impact on agriculture and the environment, Earth's ecological health and diversity remain, for many reasons, totally dependent on the biological process. And biological nitrogen fixation is the exclusive province of microbes, indeed the even more restricted province of prokaryotes—bacteria and archaea. One wonders why. We do not have a compelling answer, but, indeed, only they can do it.

We have suggested that the flow of nitrogen from the atmosphere through plants and animals and back again to the atmosphere is a cyclic process. In fact, the flow is a bit more complex; it consists of a set of cycles. Collectively, these cyclic conversions constitute the nitrogen cycle. In addition to nitrogen fixation, three other essential steps in the cycle (ammonia to nitrite, nitrite to nitrate, fixed to gaseous nitrogen) are also totally dependent on prokaryotic microbes.

The cycle would cease were there gaps anywhere in it. We will encounter these steps throughout the book.

Other bioelements, including carbon, oxygen, sulfur, and phosphorus, undergo similar sorts of cyclic conversions in nature. They, too, are essential for life. All these "cycles of matter" are poster children for microbes' many essential roles in keeping planetary ecology reasonably intact.

The inconspicuous swellings on the sweet pea's roots are a small but important part of a much grander tableau, the massive set of microbe-mediated conversions on which all life depends. Next we will encounter a completely different nitrogen-fixing symbiosis of microbes and plants.

Dark Blue-Green Spots on Liverworts

Careful examination of damp places in nature, perhaps a drippy, roadside embankment or streamside tree trunk, frequently reveals

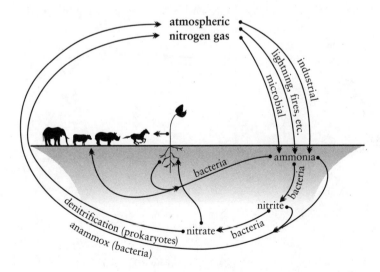

The nitrogen cycle.

the presence of liverworts or hornworts. And a careful examination of them usually reveals small (a millimeter or so in diameter) blue-green spots. These spots are packages of symbiotic cyanobacteria (once called "blue-green algae") fixing nitrogen for the plant. They can tell a fascinating story of how they got there and how they make their nitrogen-fixing contribution to the plant.

But first a bit of botany might be a useful aid to locating and recognizing hornworts and liverworts, with their delightful ancient names. "Wort" is an Old English word for "plant"; "liver" refers to the plant's shape and the presumption that it would relieve complaints of that organ. "Horn" describes the extensions that protrude from this plant's liverwort-like body. These small dwellers in damp places are bryophytes, which is Latin for "moss plant." They are nonvascular plants, lacking xylem and phloem to move water and nutrients from one part of them to another, and they do not differentiate into roots, stems, leaves, and flowers. They lack the evaporation-prevention layer (cuticle) on their surface that leaves have. That is why they are restricted to damp places; some are fully aquatic. The life cycle of hornworts alternates between two quite different forms that hardly resemble each other, one with a single

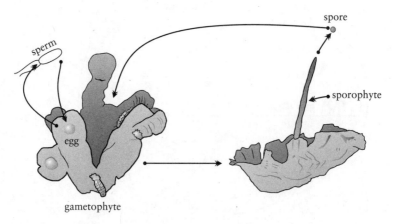

Life cycle of a hornwort.

set of chromosomes (haploid) and the other with a double set (diploid). The haploid (called a *gametophyte*) form of a hornwort is a green, flat little plant with liverlike lobes an inch or so across. It produces gametes (equivalent to our species' sperm and eggs) that fuse and develop into the diploid (or *sporophyte*) horn-shaped spikes that protrude vertically from the gametophyte and give the plant its name.

The small green spots on the lower edge of the gametophyte, small pockets in the plant that are packed with cyanobacterial cells, can be seen with the naked eye, although a hand lens is helpful. Frequently, the particular cyanobacterium is *Nostoc punctiforme*.

For the most part prokaryotes do not differentiate into different forms as most eukaryotes do. But *N. punctiforme* is a notable exception. It is marvelously skilled at developing into a variety of different forms; possibly it is the prokaryotic world record holder in this category. *N. punctiforme* can live independently in nature, but it can also seek out certain plants and form intimate symbiotic relationships with them.

When growing alone in an environment containing adequate fixed nitrogen, *N. punctiforme* exists in the form of long filaments of cells, but it differentiates into a state capable of fixing nitrogen when its supply of fixed nitrogen becomes limiting. Of course, that means making the necessary enzyme, nitrogenase. But it also changes the form and shape of certain of its cells in order to provide a low-oxygen place to use this oxygen-sensitive enzyme. And that is a special challenge for these oxygen-producing phototrophs, the cyanobacteria. They not only have to exclude atmospheric oxygen, they have to protect their nitrogenase from the oxygen that they themselves produce. *N. punctiforme* approaches the challenge as many other species of cyanobacteria do. At regular intervals in its chain of cells, individual cells differentiate to become highly specialized cells called *heterocysts*. These are nitrogen-fixing factories that supply their product to neighboring cells in the filament. Only the heterocysts make nitrogenase, which they protect from damaging oxygen. They do this by destroying the component (called *photo system two*) of their photosynthesis apparatus that generates oxygen. Thus, the nitrogenase in heterocysts is protected from internally generated oxygen. But *N. punctiforme* is normally exposed to oxygen in the air. The thick wall that they form around themselves as they differentiate takes care of that. It prevents external oxygen from entering the cell. Thus, nitrogenase in the heterocyst can function because it is protected within its oxygen-free bubble.

Sometimes environmental conditions get even bleaker than just running out of nitrogen. *N. punctiforme* meets this more severe challenge by differentiating in another way. Some of its cells go into a bunkerlike state in which they enlarge and develop thicker walls. They become highly resistant resting structures called *akinetes*. Unlike heterocysts, which lie at the dead end of a one-way differentiation, unable to become vegetative cells again, akinetes can germinate into vegetative cells when favorable conditions return.

N. punctiforme, however, undergoes its most complex and elaborate differentiation when it receives a chemical invitation from a

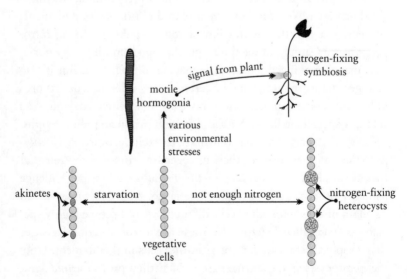

Developmental potential of a cyanobacterium, *Nostoc punctiforme.*

plant, for example a hornwort, to live together with it in a mutualistic nitrogen-fixing symbiosis. A nitrogen-deprived environment is the condition in which the plant will benefit most from the symbiosis. In such an environment, the filaments of free-living *N. punctiforme* are interspersed with nitrogen-fixing heterocysts. The plant's invitational signal causes the filaments to rupture at the junctions between heterocysts and vegetative cells, thereby forming a large number of heterocyst-free fragments of the filament. As though to indicate a sense of purpose, these fragments develop pointed ends and the capacity to move by gliding motility (movement on a solid surface without benefit of flagella by a still poorly understood mechanism). These motile fragments, called *hormogonia,* move toward the plant's attractive signal, which, in the case of the hornwort *Anthocerus punctatus,* is released from small pockets. When the hormogonia reach the hornwort, they enter the pockets and differentiate further, becoming a mass of nitrogen-fixing heterocysts. In certain respects, the *N. punctiforme–Anthocerus* symbiosis resembles that between rhizobia and legumes like the sweet pea. The plant supplies carbohydrate nutrients and the microbe supplies fixed nitrogen. But in other respects it differs significantly. *N. punctiforme* does not differentiate into a special form in the symbiotic state; it assumes the same nitrogen-fixing form symbiotically as it does when living alone.

Bryophytes are not the only plants that enter into nitrogen-fixing symbioses with cyanobacteria. Others include cycads (the palmlike gymnosperms, sometimes called "Sago palms," so prized by gardeners), herbaceous flowering plants belonging to the genus *Gunnera,* and a small floating water fern, *Azolla.* These cyanobacterial associations are widely spread through the plant kingdom. Most of them seem quite casual—just a package of cyanobacterial cells growing in a small pocket in the plant. But they are also highly specific. Only certain plants offer cyanobacteria homes for symbiosis. The specificity should not be surprising when one considers the complexity of the process of forming the symbiosis: the plant must release specific chemicals to attract cyanobacteria and to coax them into becoming nitrogen-fixing machines, and it must nourish them with food in the form of carbohydrates as the cyanobacterium nourishes the plant with nitrogen.

Another cyanobacterium, *Anabaena azolla,* in association with *Azolla,* has undoubtedly the greatest impact on human affairs. *Azolla* was named by the great early evolutionist Jean-Baptiste Lamarck. It means "dry" *[azo]* "kill" *[ollyo],* and for interesting reasons. In Southeast Asia, especially Vietnam, *Azolla* is cultivated in ponds and applied to rice paddies at the time of planting. By harvest, *Azolla* growth completely covers the surface of the paddy, providing weed control through shading. Then it is plowed into the soil, adding about twenty-five pounds of fixed nitrogen per acre. Growing *Azolla* for distribution to rice farmers has been a cottage industry in parts of Vietnam probably since the eleventh century. The technology spread throughout the country in the 1950s and contributed significantly to the increase in rice crops that occurred in the 1960s and 1970s. Although *Azolla* technology did not spread significantly to the western world, it is used in California for the production of organic rice. There, its use is completely mechanized, including sowing the fields with *Azolla* from airplanes before planting the rice.

No matter how nitrogen is fixed by a microbe, it is afterward converted to a form that plants can readily utilize. The fixed nitrogen then passes on to animals when they consume plants. Finally, as plants and animals die and decay, nitrogen is returned to the atmosphere. We will now follow some of the microbe-manipulated travels of nitrogen once it has been fixed by some prokaryote—first of all, in the manure pile.

A Manure Pile

A manure pile offers tenement housing to a vast number of microbes. We have already encountered many of its inhabitants in previous sightings: cellulolytic bacteria, such as those in the cow's rumen, which attack straw in the pile; fermenting microbes, which utilize some of the sugars that the cellulolytic bacteria release; and, most probably, archaea, which make methane from the leavings of their neighbors. But we smell something else going on. The distinct odor of ammonia (NH_3), made more intense if urine was added to the pile, signals a link in the global nitrogen cycle that we have not yet encountered.

The ammonia we smell is being released from nitrogen-containing organic compounds—a process called *ammonification*—by many kinds of microbes as they use these substances as nutrients. The ammonia rising from the pile contains the excess nitrogen that the microbes do not need. Added urine intensifies release of ammonia because it contains significant amounts of urea, which contains lots of ammonia. Microbes that produce the enzyme urease release this pungent gas from it, much as *Helicobacter pylori* does in our stomach. Urea ($(NH_2)_2-C=O$), in fact, is just a chemical package of ammonia and carbon dioxide, which urease breaks open by adding water to it.

If the manure pile is turned occasionally to introduce air into it, less ammonia is released from the pile, because much of the ammonia is nitrified: by way of two microbe-mediated transformations, the ammonia is converted to nitrate (NO_3^-) through nitrite (NO_2^-) as an intermediate. The two groups of microbes that carry out this two-step transformation are similar—both acquire metabolic energy by aerobic respiration of an inorganic compound, and they are closely related. But they attack different compounds in the pile. Members of one group, ammonia-oxidizing bacteria (AOB), typified by species of *Nitrospira*, respire the ammonia that we smell in the pile, oxidizing it to nitrite. Members of the other group, nitrite-oxidizing bacteria (NOB), typified by species of *Nitrobacter*, respire the nitrite produced by ammonia oxidizers, oxidizing it to nitrate. Together, rather like Jack Sprat and his wife, by the overall process called *nitrification*, they convert ammonia to nitrate. Thus they consume ammonia, completing one leg of the nitrogen cycle in the manure pile and elsewhere on the planet. No other organisms mediate these transformations. So without either of these two groups of nitrifying bacteria, Earth's nitrogen cycle could not function, and all other organisms could not exist.

Although distinct and highly specialized, ammonia-oxidizing and nitrite-oxidizing bacteria cooperate intimately in nature. They even live together in clusters of several thousands of cells, composed of approximately equal numbers of cells from each group. A mass of *Nitrospira* cells forms the core of the cluster, which is surrounded by an external layer of *Nitrobacter* cells at the surface. Presumably ammonia diffuses into the center of the mass and nitrate comes out. The reasons why members of these two groups cling to one another are not clear, but such intimate clustering of the two groups and resulting metabolic cooperation must be widespread because very little nitrite accumulates in nature and considerable quantities of nitrate do.

The reasons for the extreme metabolic specialization of the two groups are also obscure. Neither metabolic undertaking offers much of an energy reward. Ammonia-oxidizing bacteria must oxidize twenty-five molecules of ammonia to obtain enough metabolic energy to convert one molecule of CO_2 into precursor metabolites, or PMs. Nitrite-oxidizing bacteria must oxidize eighty molecules of nitrite to accomplish this vital task. Moreover, ammonia-oxidizing bacteria are inhibited by nitrite, their own metabolic endproduct. So in nature they are dependent on having nitrite-oxidizing bacteria nearby to use it up. We might wonder why an organism capable of oxidizing ammonia all the way to nitrate did not evolve. It would have greater energy benefits and be capable of independent existence. But this is just one of the many larger questions that come from microbe watching.

Some but not all members of each group of nitrifying bacteria can be readily cultivated by themselves in the laboratory. The requirements for their cultivation are curious, surprising, even bewildering. Of course, because they do not metabolize organic compounds and, therefore, derive all their PMs from CO_2, nitrifying bacteria do not need organic nutrients to grow. Even though they flourish in environments rich in organic nutrients, such as manure piles, they cannot tolerate the presence of even small amounts of organic nutrients when they are grown by themselves in the laboratory. Perhaps this oddity is related to why representatives of the two groups are found bundled together in nature. Living alone in the laboratory is not what they evolved to do. Probably we should not be surprised that they respond unexpectedly to such an unusual condition.

In nature, most of the nitrate, the endproduct of nitrification, is either utilized by plants as a nutrient or converted to gaseous nitrogen by other microbes. But in particularly nitrogen-rich environments, such as the turned manure pile we are viewing now, nitrate accumulates as sodium or potassium salt ($NaNO_3$ or KNO_3). Either of these salts alone or a mixture of the two goes by the name saltpeter (*peter* meaning "rock"), although the term is sometimes reserved for KNO_3.

The microbial product saltpeter has been prized since ancient times because, being a powerful oxidizing agent, it makes things burn or even explode. One of the earliest uses was as an ingredient of incense. Greeks and Romans burned incense to drive away demons; ancient Israelites burned it as part of their liturgy. Later Chinese used saltpeter to make firecrackers. It is an essential and major component of gunpowder, most of which is about 75 percent saltpeter, 15 percent sulfur, and 10 percent charcoal. When gunpowder (also known as "black powder") was first used, probably in the thirteenth century, it quickly became an essential component of warfare and remained so until it was replaced in the late nineteenth century by smokeless powder (cellulose nitrate). An ample source of microbe-produced saltpeter did not become available until the mid-nineteenth century when the massive deposits of guano-derived saltpeter from the eastern slope of the coast ranges in the Atacama, Tarapaca, and Antofagasta deserts of northern Chile were first exploited. They had been formed over long periods by nitrifying bacteria from nitrogenous material in dung that collected as birds congregated there in huge numbers.

So for about 600 years small-scale, microbe-produced saltpeter was the sole source of this essential component of gunpowder. There were several ways of producing and collecting it. Manure piles, sometimes meticulously managed, were a major source. They were irrigated with urine to increase the yield. Maintaining the aerobic conditions that nitrification needs was a challenge. One approach was to modify the pile so that it consisted mainly of loosely packed straw to which urine was added. Air could readily circulate freely through such a pile, and urine, through the activity of urease, provided an abundant source of ammonia. As we will see, aerating the pile was not just necessary for making saltpeter, it was essential for preserving the saltpeter that had already been made. When aeration was neglected, the pile became anaerobic. Then other microbes destroyed the saltpeter that had accumulated.

Saltpeter also just appeared without active human intervention. In times past when humans lived more intimately with animals, saltpeter formed on damp interior masonry walls of houses and animal shelters as a snowlike efflorescence of white crystals or scales, where it could be scraped off and collected. Its formation was considered quite mysterious. Some thought it was a cause or consequence of leprosy. But in fact, in a not-too-fastidiously-maintained structure, ammonifying bacteria could produce ammonia from nitrogenous waste material that had accumulated, and the volatile ammonia was converted to saltpeter by a consortium of ammonia-oxidizing and nitrite-oxidizing bacteria that developed on the structure's walls.

Earthen floors in barns, dwellings, and outhouses accumulated saltpeter as well. There, ammonification, ammonia oxidation, and nitrite oxidation all occurred together, converting nitrogenous waste to saltpeter. Saltpeter could be recovered from such soils by putting them into vats of water to extract the highly water-soluble saltpeter and then evaporating the water to recover the saltpeter in solid form.

Of course, governments were intensely interested in obtaining adequate supplies of saltpeter to meet their military purposes and mandated a variety of ways to reach their goals. England's parliament commissioned "saltpeter men" to collect nitrogenous soil from property owners and extract saltpeter from it. The owners were required by law to cooperate. Government representatives in France were given the authority to confiscate soil from stables and outhouses. In Sweden up until 1835, landowners were required to deliver a prescribed quantity of saltpeter to the state.

Nitrate is also a component of explosives other than gunpowder. One that we became painfully aware of after the Oklahoma City terrorist attack in 1995 that destroyed the Alfred P. Murrah Federal Building is ammonium nitrate (NH_4NO_3), a fertilizer made of ammonia and nitrate. The perpetrators at Oklahoma City added diesel to their two and a half–ton truckload of ammonium nitrate to make the mixture more readily explosive and powerful. Ammonium nitrate is also a component of cluster bombs. The fertilizer alone is highly explosive, although difficult to detonate. But inadvertent explosions have happened many times with devastating consequences. Detonation sometimes occurred in the past when piles of solidified ammonium nitrate were naively blasted to loosen them. Major explosions caused by fire or even collision continue to the present. Fires on ships, trucks, and trains carrying fertilizer have caused horrendous explosions.

Of course, nitrifying bacteria make more important contributions to the world than merely supplying saltpeter for gunpowder. They provide the sole route by which nitrate is made in nature and hence supply one of the main starting points for replenishing nitrogen in the atmosphere. Replenishment is an ecological necessity because atmospheric nitrogen is constantly being used in the process of nitrogen fixation.

Nitrifying bacteria are abundant and highly active in most soils. Confirmation of this is the widespread agricultural use of gaseous (anhydrous) ammonia as a fertilizer. Ammonia is introduced into the soil and covered. It is quickly converted by nitrifying bacteria to nitrate, the form of nitrogen that plants most readily use. The lowly, smelly manure pile is a microcosm of monumental global microbial events. Next, in a larger, more industrial version of it I describe the completion of the nitrogen cycle.

A Wastewater Treatment Plant

Treating wastewater or sewage effectively is a relatively modern innovation, coming into widespread use only during the second half of the twentieth century. Before that, most municipalities did not treat sewage. They simply dumped it into the most conveniently available body of water—a river, a bay, or the ocean—and hoped for the best. The consequences were always unpleasant and sometimes disastrous. When large amounts of sewage were added, the receiving water became polluted, and aerobic microbes used up all available oxygen. Metabolizing their feast of organic nutrients, they turned the water anaerobic, killing all oxygen-requiring aquatic organisms, including fish. Then, anaerobic microbes took over, producing their complement of stinky endproducts of fermentation and anaerobic respiration. Sulfate-reducing bacteria turned marine dumping places black. On occasion the transition to anaerobiosis was sudden.

Possibly the most dramatic, certainly the most notorious, of these transitions was the "Big Stink" that occurred in the Thames River adjoining London in the summer of 1855—an odoriferous microbial sighting we might all be grateful to have missed. A number of factors cooperated to precipitate it. Replacing chamber pots with newly developed flush toilets massively increased the total liquid volume of sewage, which overwhelmed London's estimated 200,000 cesspits, where chamber pots had been emptied. The cesspits overflowed into the street drains, which had been constructed principally to accommodate the outflows from factories and slaughterhouses, and the combined load of waste flowed into the Thames. It was a hot summer, which accelerated the metabolism of aerobic bacteria acting on the waste, rapidly using up any dissolved oxygen.

The Thames suddenly went anaerobic, producing the horrendous "Big Stink," as it was called. In an attempt to moderate the odor, the House of Commons soaked its curtains in lime and considered moving out of town. Law courts made plans to relocate to Oxford. Later in the summer, heavy rains provided temporary relief, by cooling London and flushing the Thames.

The primary aim of modern wastewater treatment is to induce microbes to reduce the organic content of sewage, its biological oxygen demand (BOD), so that its outflow will not deplete the receiving body of water's oxygen supply. First they must mechanically separate larger objects from incoming sewage, which is known as the "primary treatment."

There are many ways that wastewater treatment facilities encourage microbes to decrease the BOD of sewage. Sometimes, they begin by employing an anaerobic digester, a large closed chamber in which anaerobic microbes convert some of the organic material to methane through cascades of conversions we have encountered in a previous chapter. The effluent methane is either torched or used as a fuel for the facility or the municipality. Undigested sludge, which accumulates at the bottom of the tank, can be dried and disposed of. Anaerobic waste digestion is also sometimes referred to as a primary treatment. We will return to the anaerobic digester after taking a brief look at the all-important aerobic component of sewage treatment. Such an anaerobic digester in the Netherlands was the venue for the relatively recent discovery of a major new microbial component of the global nitrogen cycle.

The liquid sewage, whether or not it has passed through an anaerobic digester, is treated aerobically to decrease the sewage's BOD by schemes designed to expose the sewage to a dense population of aerobic microbes with abundant access to air. Any of a variety of installations will work. The most prevalent, perhaps, is a trickling filter, which is simply a bed of rocks several feet deep, large enough to provide air channels through them. Sewage is sprayed from above, often from a rotating arm, and treated sewage trickles out from the bottom. Soon the rocks in the filter become covered with a layer of slime, a biofilm densely packed with aerobic respiring bacteria, making the trickling filter a remarkably efficient and rapid oxidizer of organic matter.

Effluent from the trickling filter is said to have undergone secondary sewage treatment. If the secondary treatment procedure is working well, this effluent contains very little organic matter; its BOD is quite low. So aerobic microbes in the receiving environment will not make it anaerobic and smelly. However, such secondary-treated sewage can cause other environmental insults, some of them major.

Because the growth of photosynthetic microbes, algae and cyanobacteria, in most natural aquatic environments is limited by the availability of a source of nitrogen or phosphorus, addition of secondary-treated sewage, which is rich in inorganic forms of both nitrogen (largely as nitrate) and phosphorus (as phosphate), to a pristine body of water stimulates voluminous growth of photosynthetic microbes, a process ecologists call *eutrophication*. So eutrophication, to a certain extent, reverses the benefits of sewage treatment. It adds organic material in the form of microbial cells, thereby increasing BOD. If the growth of photosynthetic microbes is profuse (causing a "bloom"), the receiving body of water can become anaerobic as aerobic microbes metabolize the remnants of dying phototrophs, just as it would if the untreated sewage had been added directly.

In order to avoid the consequences of eutrophication, tertiary treatments designed to remove nitrogen and phosphorus can be incorporated into sewage treatment. Removal of both nitrogen and phosphorus can be accomplished chemically, but microbial removal offers distinct advantages.

Phosphorus, which is present in the form of phosphate (PO_4^{3-}), can be taken up by certain bacteria such as species of *Acinetobacter*, which are particularly adept at polymerizing phosphate and storing the resulting macromolecule, polyphosphate, inside their cells as solid granules, when air nutrients and phosphate are available. Under ideal conditions *Acinetobacter* species can store up to 80 percent of the weight as polyphosphate. Meticulously managed, such bacteria can act as a phosphate-scavenging pump, removing phosphate from sewage and releasing it in relatively pure form, possibly to be used commercially. Under the high-volume, practical conditions reigning in a sewage plant such phosphate-accumulating bacteria can and are used to remove phosphate from sewage. Environmental engineers call the process "enhanced biological phosphate removal" (EBPR). It involves passing treated sewage with a small BOD successively through two tanks, the first anaerobic, the second aerobic, and recycling a portion of the bacteria-containing sludge from the second tank into the first. The rest of the sludge is removed along with the solid polyphosphate it contains. Phosphate scavenging and removal occurs in the second tank. No additions of appropriate bacterial strains are necessary. A population of effective phosphate-removing bacteria develops simply by the cycling between the anaerobic and aerobic tanks. Because it has not yet been possible to cultivate the bacterium or bacteria that flourish in these tanks, it is called *Candidatus Accumulibacter*. (*Candidatus* is a prefix given to a microbe that has yet to be cultured.)

Nitrate can also be removed in tertiary sewage treatment by encouraging bacterial denitrification. Denitrification is a serial cascade of anaerobic respirations through which nitrate is converted to nitrogen gas. Sequentially, nitrate (NO_3^-) is converted to nitrite (NO_2^-), which in turn becomes nitrous oxide (N_2O), then nitric oxide (NO), and finally nitrogen gas (N_2). Each of the chemical

members in the cascade serves as an oxygen substitute in an anaerobic respiration that produces the next one. For example, organic
nutrients and nitrate produce carbon dioxide and nitrite, releasing
metabolic energy.

The nitrate, which is the starting point of denitrification, is
formed during secondary sewage treatment when the nitrogenous
components of sewage are converted successively by ammonification and nitrification to nitrate, just as they are in a manure pile.

Tertiary treatment to remove this nitrate involves fostering denitrification. It is easy enough to do. It is only necessary to establish anaerobic conditions and provide some untreated sewage as a
source of organic nutrients. That is why it was so important for
saltpeter makers to keep their manure piles constantly aerated. As
soon as the pile became anaerobic, denitrification would begin to
use up their valuable saltpeter.

Denitrification completes the nitrogen cycle, which as you will
recall begins with microbial nitrogen fixation, produced through
such means as the rhizobia in the nodules on the roots of leguminous plants or the dark blue-green spots on hornworts. Farmers
often revile denitrification because it removes from the soil the fixed
nitrogen they purchased as fertilizer, thereby diminishing the soil's
fertility. However, returning nitrogen to the atmosphere is vital to
life's continuation on Earth. Were nitrogen not to return to the
atmosphere, Earth's total supply would accumulate as nitrate, a
highly water-soluble and mobile form. Nitrate does not bind to soil
as other forms of nitrogen such as ammonia do. It would eventually all end up in the oceans, leaving the land masses as nitrogen-
depleted deserts. The return of nitrogen to the atmosphere must be
a near-continuous process because the atmosphere's supply of nitrogen is not particularly large compared with the rate at which it is
used. The half life of atmospheric nitrogen, which can be calculated
from its steady-state values in the atmosphere and the rate at which
it is utilized (or resupplied), is only about 15 million years—a moment in the context of Earth's 3.5-billion-year biological history.

The process of bacterial denitrification (one archaeon is also known to denitrify) was discovered in 1886 and considered for more than a hundred years to be the only route by which atmospheric nitrogen is replenished. Then, in the 1990s, while studying the mysterious loss of ammonia from an anaerobic digester at a wastewater treatment plant in Holland, J. G. Kuenen and M. S. M. Jetten discovered a totally different process by which certain bacteria convert fixed nitrogen into its gaseous atmospheric form. Recall that the major known route of utilizing ammonia (NH_3) is its oxidation by nitrifying bacteria, which is what happens in a manure pile. No oxygen, no denitrification. The Dutch microbiologists discovered that nitrite (NO_2^-) was disappearing from their reactor along with the ammonia. They guessed that the loss of the two forms of nitrogen might be related—that one might be reacting with the other to form nitrogen gas.

To test their hypothesis they added to the digester ammonia that was composed of the rare heavy nitrogen isotope (^{15}N). The consequences proved that indeed they had guessed right. Nitrogen gas (N_2) with one atom of the heavy isotope (^{15}N) and one light one (^{14}N) evolved from the digester. Without doubt the heavy one came from the added ^{15}N-ammonia and the ordinary light one came from the nitrite formed in the digester. The bacterium that causes this reaction, which has so far resisted being cultivated in pure culture, has been assigned the tentative name *Candidatus Brocadia anammoxidans*. *Candidatus B. anammoxidans* uses this way of oxidizing ammonium at the expense of nitrite, which has been termed *anammox* (for *an*aerobic *amm*onia *ox*idation), to generate its metabolic energy. Because only inorganic compounds participate in the anammox reaction that *B. anammoxidans* exploits to derive its metabolic energy, it uses CO_2 as a source of carbon to make its PMs. So *B. anammoxidans,* as well as similar bacteria that live by anammox, is in the class of microbes called *chemoautotrophs.* "Chemo" refers to its exclusive use of inorganic compounds to generate metabolic energy, and "autotroph," meaning self-feeding, refers to its use of CO_2 as a source of PMs.

Subsequently it was discovered that anammox bacteria are not

just a peculiarity of anaerobic digesters in wastewater treatment facilities. They are widespread in nature and play a major role in the
world's nitrogen cycle. In certain environments such as marine sedi-
ments these bacteria may account for as much as two-thirds of the
flow of fixed nitrogen back to the atmosphere.

It might seem odd for anaerobes to be dependent on nitrite,
which is produced aerobically by ammonia-oxidizing bacteria.
Most probably, however, anammox and ammonia-oxidizing bacteria live intimately and cooperatively near the surface layers of the
silt at the bottom of bodies of marine and fresh water. Ammonia
is generated through ammonification of nitrogenous organic compounds in the silt, and it is oxidized by ammonia-oxidizing bacteria
in the aerobic environment at the silt's surface or just above it. In
the anaerobic zone just a bit lower, anammox bacteria convert ammonia and nitrite to nitrogen gas.

This aerobic-anaerobic interplay between ammonia-oxidizing
and anammox bacteria is probably ecologically analogous to the
close association between fully aerobic ammonia- and nitrite-
oxidizing bacteria we witnessed in the manure pile. It offers the
same advantage to the ammonia oxidizers, namely removing the
nitrite they produce, which would otherwise be toxic to them.

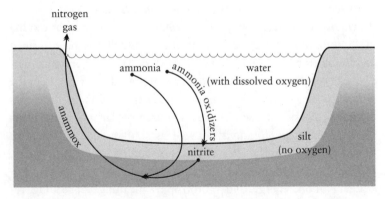

Natural history of anammox.

Wastewater treatment facilities might never become major tourist attractions, but they do contribute monumentally to maintaining our people-dense world and they are quite special centers of interest for microbe watchers. The slimy microbial coating on the rocks in the trickling filter is a microcosm for much of the world's carbon and nitrogen cycles; tertiary treatment and the denitrification complete Earth's vital nitrogen cycle. Except for its nitrogen-fixation component, the entire nitrogen cycle can be viewed in a wastewater treatment plant. And we should not forget the major intellectual contribution of the anaerobic digester. It was the site of a discovery that filled a hundred-year lacuna in our knowledge of our planet, namely, how a vast majority—80 percent—of its atmosphere was formed and maintained. Such discoveries remind us how incomplete is our knowledge of microbes' contributions to Earth's ecology.

6

Cycling Sulfur

Various chemical forms of an essential bioelement, sulfur, cycle through our environment in a flow reminiscent of the cycling of nitrogen. This sulfur cycle is as requisite to the continuation of life on Earth as the nitrogen cycle is, but it differs fundamentally in two respects: it does not pass through a massive atmospheric reservoir and it does not contribute in a major way to the constituents of cells. Still, sulfur is a necessary component of all cells, being present in proteins and RNA. For example, one of our essential amino acids is sulfur-containing methionine, which we must acquire from eating microbes or plants, although it might pass through the tissues of animals on the way.

Downstream from an Abandoned Gold Mine

This colorful, strangely beautiful, and highly visual microbe sighting is fortunately becoming less common. A stream with a reddish-orange bottom, called *yellow boy,* is the downstream consequence of the nasty, microbe-enhanced pollution of the water flowing from a mine. The culprit microbes, like anammox bacteria, live by oxidizing an inorganic compound, in this case a reduced sulfur compound, and using CO_2 as the source of PMs. As they oxidize the reduced sulfur, they solubilize iron pyrite (FeS_2) in the mine. These microbes belong to a class called *acidophiles* (acid lovers) because they flourish in an extremely acidic environment—in this case, one that they created themselves. When the mine water flows into a stream, its acidity is partially neutralized, causing the solubilized iron to precipitate as reddish-orange iron hydroxide, forming yellow boy.

The world's most horrific example of this pollution occurs in the Rio Tinto (meaning "dyed river" in Spanish and implying "red river"). Because the receiving water does not have sufficient neutralizing capacity, the entire river remains acidic and red. The Rio Tinto and its estuary adjoin Huelva (where Columbus set sail on his first voyage to the New World) in southwestern Spain. The river drains mines in the mineral-rich, massive Iberian Pyrite Belt, which is 150 miles long, eighteen to twenty-four miles wide, and thousands of feet thick. Mines have operated there for approximately 5,000 years, successively by the Iberians, Phoenicians, Romans, Visigoths, and Moors. Then after a pause of a few hundred years, operations were begun again by the British in the nineteenth century and continued until the late twentieth century. Copper production ceased in 1986, and gold and silver stopped in 1998.

The mines in the Iberian Pyrite Belt have a long and romantic history. Some thought they were the legendary mines of King Solomon. Indeed, a section of the belt is still known as Cerro (hill) Salomón. The yields of these mines had great impact on human history. Romans made gold and silver coins from metal mined here, and the mines' bounty ushered in the Copper and Bronze Ages. The mines were open pits. As individual mines grew, they merged, forming huge, miles-wide craters. Estimates set the total amount of material removed over the centuries at 1.6 trillion metric tons. Because the pits extended below the water table, they flooded and water drained out. Metabolism of the sulfur-oxidizing bacteria acidified the water and this acidity solubilized heavy metals. The result, even today, is a horrendous toxic mixture. The Rio Tinto system is probably the most polluted estuarine region in the world.

There are numerous others throughout the world, including sites in Britain, France, Canada, and the United States, where mining and microbes have collaborated to pollute huge amounts of water. An abandoned mine in California, Iron Mountain Mine, is certainly one of the most notorious. My uncle, Bert Stewart, a conservative guy who never creased his fedora hat because it would wear out more quickly, worked there as a mining engineer. I visited it as a child. It was a mess even then. Long recognized as one of America's most toxic sites, it has been listed as a federal Superfund site since 1983. Chinook salmon kills in the Sacramento River caused by its drainage were reported as early as 1899. The mine holds the record for containing the most highly acidic water on Earth. Samples of its waters taken in 1990 and 1991 had a pH value of −3.6, over a thousand times more acidic than battery acid. The extreme acidity of the water is a consequence of the combined effects of the bacteria and evaporation.

Iron Mountain Mine began operating in the 1860s, producing iron, silver, gold, copper, zinc, and pyrite. Pyrite, iron sulfide, is sometimes called "fool's gold" because of its shiny metallic look. It is used industrially as a source of sulfur. Its name comes from the Greek word *pura* (meaning "fire") because of the sparks it makes when struck against steel. It was used in early firearms, such as the wheel lock, to ignite gunpowder. Operations at Iron Mountain ceased in 1963, but its power to pollute goes on because of mining's gigantic assault on the mountain. The Mountain Copper Company, Ltd., mined it simultaneously inside and out in several ways: underground using the open stope (pit with ledges) method, on the surface in an open pit, and on its sides by side-hill mining. These collective attacks on the mountain's integrity caused the entire mountain to split open, exposing its pyrite contents to rain and oxygen. This was all the microbes needed to start their own on-slaught.

The principal microbial attacker of the exposed pyrite is *Acidithiobacillus ferrooxidans,* the sulfur-oxidizing, acidophilic bacterium I mentioned earlier. In order to acquire metabolic energy, *A. ferrooxidans* oxidizes both components of pyrite, its reduced iron (ferrous ion, Fe^{2+}) as well as its reduced sulfur (sulfide ion, S^{2-}) via aerobic respiration. Like all microbes that live by respiring inorganic compounds, it makes its PMs from CO_2, so the completely mineral diet supplied by the mountain is sufficient to sustain *A. ferrooxidans.* It respires the ferrous ion, producing ferric ion, water, and metabolic energy, and it respires sulfide ion, producing sulfuric acid and metabolic energy.

Oxidation of sulfide ion is the powerful generator of acidity. The sulfuric acid it produces lowers the pH of the drainage water to levels at which toxic minerals solubilize, including copper, cadmium, zinc, and other heavy metals. Superfund remediation efforts have now decreased the copper and zinc content of the drainage water from Iron Mountain by 90 percent and significantly raised its pH.

The interior chambers of the abandoned copper and zinc mine at Iron Mountain remain a fearsome environment with temperatures over 120°F and a ceiling decorated with brightly colored stalactites dripping extremely acid, heavy metal–laced water.

This type of microbe-assisted pollution is called *acid mine drainage.* Mining for metals in pyrite-rich regions is not the only source of it. Many coal mines contain sufficient reduced sulfur to produce acid drainage by essentially the same process. They do not, however, form yellow boy because they contain very little iron, but they do cause extensive ecological damage because of their acidity.

Collectively, the environmental threat of acid mine drainage is enormous. The U.S. Environmental Protection Agency (EPA) estimates that there are between 200,000 and 550,000 abandoned mine sites in the United States, some of which are in quite remote places. Cost estimates for cleaning the worst of them vary between 7 and 72 billion dollars.

The approaches, many of which have been successful, to detoxifying microbe-caused acid mine drainage are also principally microbe based, although some do rely on chemically neutralizing the acidity. The remediating bacteria are acid-tolerant representatives of the sulfate-reducing bacteria that we discussed in an earlier chapter—those that inhabit the Black Sea and marine mudflats. Recall that these bacteria acquire their metabolic energy by exploiting an anaerobic respiration of organic nutrients while reducing sulfate to hydrogen sulfide. In the case of acid mine drainage, the sulfate is present as sulfuric acid. The respiration of organic nutrients with sulfuric acid (H_2SO_4) produces hydrogen sulfide (H_2S), CO_2, and metabolic energy.

By reducing the sulfuric acid, the sulfate-reducing microbes accomplish the first imperative of remediation: decreasing the acidity of the acid mine drainage. Producing hydrogen sulfide as a product accomplishes the second: removing toxic metals. Hydrogen sulfide causes cadmium, copper, lead, nickel, and zinc to precipitate as metal sulfides. In addition, it removes arsenic, antimony, and molybdenum by forming more complex sulfide minerals.

In order to foster remediation, it is only necessary to provide an organic nutrient and a venue for it to occur—no need to add sulfate-reducing bacteria. They are already present in the environment. In some cases, the venue is a pond, constructed especially for the purpose; organic nutrients are added to it. In others, remediation occurs in the mine itself, and the organic nutrients are added through new shafts drilled into it. A number of organic nutrients have been evaluated and shown to be effective. Some are nutrients that the sulfate-reducing bacteria can use directly; methanol and ethanol are examples. But these are relatively expensive. Cheaper cellulosic materials, such as manure and straw (with added nitrogen), which sulfate-reducing bacteria cannot use directly, are just about as effective. Their utilization, however, depends on other microbes that carry out a cascade of microbial metabolisms, somewhat reminiscent of those that occur in a belching cow. They convert the straw and manure into nutrients that the sulfate-reducing bacteria can use. Cellulose-decomposing and fermentative microbes convert these cellulosic materials into organic acids (lactic and acetic, for example) that the sulfate-reducing bacteria can use as organic nutrients. With all the microbial activity that takes place, the remedial environment rapidly becomes anaerobic and begins to resemble a cow's rumen or the bottom of a quiet pond, a suitable place for methane-producing bacteria. Methane is, indeed, produced during remediation.

The most unwelcome pyrite-decomposing bacterium, *Acidithiobacillus ferrooxidans,* is the root cause of pollution resulting from acid mine drainage. But its growth is also fostered in certain mining operations to recover metals from low-grade ores. In an uncomplicated process called *biological heap leaching,* a pile of low-grade copper ore, in which the copper is bound in a sulfide matrix, is irrigated. *A. ferrooxidans* grows at the expense of the ore's sulfide, as it does on pyrite, breaking down the matrix and releasing its copper, which flows from the pile into a catch basin as a blue solution of copper salts. Copper metal is recovered from the pond and the remaining acid solution is recycled by using it to irrigate the pile. Fully 25 percent of the world's copper, worth billions of dollars annually, comes from this microbe-mediated process. A similar process is used to remove sulfide from low-grade gold ores before they are extracted with cyanide (a commonly used method of recovering gold). This microbial process is an alternative to smelting, roasting the ore at high temperature to drive off the sulfur. The microbial treatment is less expensive and more effective, in one case having increased recovery from 70 to 95 percent. And the microbial process is far less devastating to the environment. Smelting releases huge quantities of sulfur dioxide (SO_2), which kills plants. Kennett, California, which now lies submerged under Lake Shasta, served as a dramatic example. When four copper smelters were operating there in the early 1900s, nearly all vegetation was destroyed within a fifteen-mile radius.

Because they are able to proliferate in such a highly acidic environment, *Acidithiobacillus ferrooxidans* and other acidophiles can be added to the list of microbes called *extremophiles*, microbes that flourish in highly hostile environments.

Like other extremophiles, acidophiles have evolved special physiological mechanisms to adapt to an environment that would be lethal to most organisms. Because the prevailing level of acidity within a cell affects almost all of its functions and components, including the stability of its proteins, organisms evolve to operate at a particular and fixed value of intracellular pH. In multicellular forms such as plants and animals, the cell's external environment, for example blood, is maintained at the same value as the cell's interior. Microbes, of course, are not afforded such luxury. They are subjected to pH environments that differ, sometimes markedly, from the value of their internal pH. They are able to adapt to different, often changing external pH environments, although very few microbes can adapt to the extreme acidity of the acidophilic environments that are encountered in some mines.

The well-studied *Escherichia coli,* for example, can thrive in environments with pH values that range from 6.0 to 8.0 (a hundred-fold difference in acidity). It tolerates values half a pH unit above and below that range. In spite of a varying external pH, it maintains its internal pH at a constant value slightly above neutrality. It does so by pumping protons (hydrogen ions) out of the cell. Of course, the concentration of hydrogen ions (H^+) is what determines pH. Acidophiles are faced with a much greater challenge, but they employ the same general strategy: they maintain an almost constant internal value of pH by pumping protons out of their cells. They maintain a somewhat lower internal pH, about 6.0 in most cases.

You may notice that *Acidithiobacillus ferrooxidans* and acidophilic sulfate-reducing bacteria form a closed loop of sulfur metabolism. The former oxidizes reduced sulfur (in the form of pyrite) to sulfate (in the form of sulfuric acid) by aerobic respiration, and sulfate-reducing bacteria return it to its reduced state (in the form of hydrogen sulfide) by anaerobic respiration. Together they form a simplified version of a sulfur cycle.

Piles of Sulfur on the Gulf Coast

Once there were huge piles of bright yellow sulfur on the Texas and Louisiana Gulf Coast. The size of these piles measured in hundreds of feet. The origin of the almost pure (99.5 percent) sulfur they contained was microbial and ancient, having been formed in the Permian Age some 250 million years ago. The microbial events that formed the huge reserves from which this sulfur was mined still make elemental sulfur in various locations on Earth today.

But the piles of sulfur themselves have largely disappeared from the Gulf Coast. They have been replaced by similar piles near installations where petroleum, gas, and oil are processed to rid them of contaminating sulfur, which when burned would otherwise pollute the atmosphere with sulfuric acid and cause acid rain, among other environmental damage. Some of the largest of these new piles are located in Alberta, Canada, where much natural gas is treated. This major new source of sulfur, arising from environmental concerns and legislation, rendered sulfur mining noncompetitive. The last mine on the Gulf Coast closed in 2000. Huge quantities of sulfur remain there unmined.

The sulfur in these underground deposits was produced by sulfate-reducing bacteria, the same anaerobic respiring bacteria in the Black Sea or the mine stream in this chapter. The source of organic nutrients they used to reduce sulfate was most probably petroleum, which is abundant in the area. The sulfate was supplied from the seawater that infiltrated the deposits. But sulfate-reducing bacteria, as you will recall, produce as an endproduct hydrogen sulfide (H_2S), not sulfur (S^0).

Indeed, sulfur is not even an intermediate in the pathway through which sulfate passes as it is reduced to hydrogen sulfide. Sulfur in the deposits is presumed to have formed as the hydrogen sulfide trapped under salt domes, which are abundant in the region, was slowly oxidized by a nonbiological chemical reaction. When hydrogen sulfide is combined with oxygen, sulfur and water result. To support this theory, the sulfur deposits are found under the salt domes.

The conviction that these sulfur deposits were, indeed, formed by microbes rests on more than mere plausibility. It is supported by physical evidence. Sulfur in nature is a mixture of four isotopes—forms of an element with different atomic weights: ^{32}S, ^{33}S, ^{34}S, and ^{36}S. Microbes are able to discriminate among these various forms. As has been shown in laboratory studies, sulfate-reducing bacteria prefer ^{32}S to ^{34}S. They use the former more rapidly than the latter. Thus, the ratio of ^{32}S to ^{34}S in the hydrogen sulfide they produce is greater than that of the sulfate they start with by several percent. And that same difference is found between the sulfur in the deposits and the sulfate in the surrounding rocks (residue from the former sea). The ^{32}S to ^{34}S ratio in the sulfur is, indeed, several percent greater—strong evidence for its microbial origin.

As noted, microbes are actively making elemental sulfur in many places on Earth today. One of the most dramatic and microbiologically explicit is occurring in lakes in the Cyrenaican area of Northeast Libya, in North Africa. The silt at the bottoms of the lakes that have been studied, Ain ez Zauia, Ain el Rabaiba and Ain el Braghi, is about half elemental sulfur, and the lakes give off a strong sulfurous odor. These lakes are relatively small, only about 100 to 200 feet wide and about five feet deep. They are fed from warm springs that supply water rich in sulfate.

The lakes' warm feed water creates a temperature gradient from top to bottom that stabilizes the lakes. Like the Black Sea, they do not turn over: they are meromictic and much of their contents are anoxic—an ideal habitat for sulfate-reducing bacteria, which are abundant in these lakes. These bacteria reduce the sulfate to the hydrogen sulfide that produces the lake's rotten-egg odor. Unlike the fate of the hydrogen sulfide produced by sulfate-reducing bacteria in the salt domes, most of the hydrogen sulfide produced in the Cyrenaican lakes is utilized by the type of anaerobic photosynthetic bacteria (purple and green sulfur bacteria) that occur in Yellowstone National Park. These bacteria form a carpetlike microbial mat, red and gelatinous, that extends several yards out from the lakes' shores. The purple sulfur bacteria are on top. They give the mat its red color. The underside of the mat is green and black because green sulfur bacteria grow there. Both of these two groups of anaerobic bacterial photosynthesizers oxidize hydrogen sulfide, producing elemental sulfur for the same metabolic purpose that plants and cyanobacteria oxidize water and produce oxygen. Sulfur granules accumulate within cells (in the case of purple sulfur bacteria) or outside the cells (in the case of green sulfur bacteria) of these

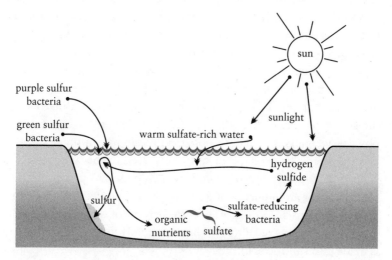

Flow of sulfur through Cyrenaican lakes.

phototrophs and later collect in the silt at the lake bottoms as bacterial cells die and disintegrate. The organic constituents of the dead phototroph's cells are not wasted. They serve as the organic nutrients for the sulfate-reducing bacteria. So the two different types of microbes that live in these lakes supply each other with essential nutrients. The sulfate-reducing bacteria supply the phototrophs with hydrogen sulfide. And the phototrophs supply the sulfate-reducing bacteria with organic nutrients.

A variation on the pattern by which sulfur forms in nature occurs in certain lakes, including Lake Sernoye in Russia. The hydrogen sulfide produced in the lower anoxic regions of that lake is oxidized to sulfur by aerobic chemoautotrophic bacteria in the upper oxygen-containing regions near the surface. We will encounter these bacteria again in this book near the deep-ocean rifts.

So we have seen three ways that vast quantities of sulfur are made in nature. In all cases, the first step is the reduction of sulfate by sulfate-reducing bacteria to hydrogen sulfide. The variation occurs in the way that hydrogen sulfide is then oxidized to sulfur. Off the Gulf Coast, the oxidation is abiotic: oxygen dissolved in the seawater simply reacts, albeit slowly, with the hydrogen sulfide. In the Cyrenaican lakes, it is oxidized by anaerobic phototrophic bacteria, and in Lake Sernoye by aerobic sulfur-oxidizing bacteria.

Sulfur has been an important article of commerce for over 2,000 years. Egyptians used it to bleach cloth and make pigments. Greeks used it to clean, and Romans used it as a pharmaceutical. Demand for sulfur increased dramatically after the Chinese invented gunpowder, which is, as we have noted, a mixture of sulfur, charcoal, and saltpeter. The demand for sulfur increased further with the advent of the industrial revolution. It was then used to make sulfuric acid (H_2SO_4), an important industrial starting material for making myriad chemicals. At present, over 90 percent of the more than 13 million metric tons of sulfur produced annually in the United States is made into sulfuric acid.

Since ancient times when sulfur was known as brimstone when burning, a symbol of God's wrath, Sicily has been a major source of the world's sulfur. This moderate-sized island off the toe of the Italian boot had a near monopoly on its production. The Sicilian process of recovering sulfur was as simple and direct as it was effective. Sulfur-laden rocks were stacked on a hillside, covered with previously recovered sulfur, and ignited. The fire melted the sulfur within the rocks, which ran out the bottom of the pile and down the hill. When it solidified, it was gathered in buckets. If purer sulfur was needed, the collected sulfur was then distilled. Powerful physical evidence, namely the ratio of ^{32}S to ^{34}C in Sicilian sulfur, proves that it, too, was made by sulfate-reducing bacteria a little more than 5 million years ago in the late Miocene age. Elemental sulfur also spews from volcanoes, but, in spite of Italy's being an active volcanic region, Sicilian sulfur is not volcanic in origin.

Mining the microbially formed sulfur on the Gulf Coast was technically much more complicated than getting it from rocks in Sicily. Conventional mining was infeasible because of the depth (about a mile) of the deposits and the quicksandlike material covering them. Herman Frasch, a nineteenth-century petroleum engineer from Germany, solved the problem in 1891 by developing and patenting a novel method of extraction, which became known as the Frasch process. A hole large enough to accommodate three concentric pipes is drilled through the salt dome and into the sulfur deposit. Superheated water is pumped down the outer pipe, melting the sulfur below, and heated air, which is pumped down the central pipe, forces liquid sulfur to gush up as a foam through the middle pipe onto a pile at the surface, where it solidifies. The economic feasibility of the Frasch process and the vast supplies of sulfur in the deposits on the Gulf Coast spelled the death of the Sicilian sulfur industry when this new mining method was introduced in the early twentieth century. The Frasch process reigned supreme for about a hundred years (although, as we saw earlier in this chapter, some sulfur was made from mined iron pyrite). Then in the early twenty-first century, sulfur derived from purifying gas and petroleum for environmental reasons took over.

In this chapter we have seen most of the transformations of sulfur that take place in nature. Like nitrogen, many of them are also cyclic.

Only two of the transformations that make up the sulfur cycle are still missing from our account—the process by which sulfur enters the components of living things (assimilation) and the step by which microbes extract it again from these components (desulfurylation), although we do get a whiff of desulfurylation whenever we smell a rotten egg (hydrogen sulfide). Sulfur is an essential component of some of the amino acids in proteins and some of the bases in RNA. Plants make these essentials from sulfate, and animals ac-

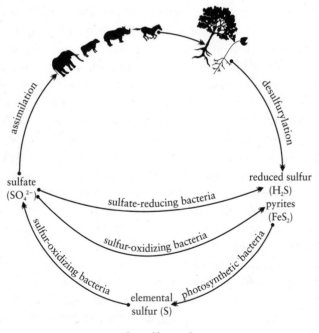

The sulfur cycle.

quire them directly by consuming plants or indirectly by consuming other animals that consumed plants. Because microbes also must have sulfur-containing protein and RNA, they too assimilate sulfur. They do so largely by the same route as plants do.

Except for assimilation and the passage of sulfur compounds from plants to animals, all steps of the sulfur cycle are the sole responsibility of microbes. The sulfur cycle is really two cycles: a large one, which involves the participation of plants and animals, and another, smaller one—the cycling of sulfur between hydrogen sulfide and elemental sulfur—that is totally microbe based. The sulfate-to-hydrogen sulfide flow in the small cycle occurs only in the absence of oxygen and is driven by energy obtained from organic nutrients. The opposite flow (from hydrogen sulfide to sulfate) depends on the presence of either oxygen or light. The complete small cycle undoubtedly occurs in the Libyan lakes because some hydrogen sulfide must be oxidized to sulfate by aerobic sulfide-oxidizing bacteria in the lakes' uppermost regions. But sulfate-reducing bacteria in the lake are not dependent on this regeneration of sulfate because more of it is constantly supplied by the springs that feed the lake. With this continuous resupplying of sulfate, the huge quantity of sulfur being trapped as elemental sulfur in the silt at the bottom of the lake does not deprive the various other sulfur-dependent microbes in the lake of the particular forms of sulfur that they must have.

The sulfur deposits on the Gulf Coast that date from the Permian Age and the sulfur cycling in Libya's Cyrenaican lakes as well as a different type in Russia's Lake Sernoye provide a glimpse of Earth's vast sulfur cycle. All of the individual conversions comprising this cycle occur in many other environments. Many should be the basis for microbe sightings, whether of sulfate-reducing bacteria or anaerobic bacterial phototrophs.

7

Cycling Carbon

Other than hydrogen and oxygen, carbon is the most abundant element in living things. It forms the backbone of organic compounds and the basis for life on Earth. Organic means carbon. Like nitrogen and sulfur, oxygen cycles through various forms and reservoirs in the environment, including the atmosphere. Its major atmospheric form, carbon dioxide, makes up a much smaller fraction of the atmosphere (0.03 percent) than does nitrogen gas (78 percent), and it cycles more rapidly. Without resupply, atmospheric carbon dioxide would last only about twenty years.

The major loop of the carbon cycle is utilization and replenishment of this atmospheric carbon dioxide. Autotrophs, principally photosynthetic organisms that include phytoplankton and plants as well as microbes that live by oxidizing inorganic compounds while using carbon dioxide to make their PMs, draw from the reservoir. And until relatively recently in Earth's history, respiration was the major route of replenishment. Depletion and replenishment occurred at approximately the same rate. The cycle was nicely balanced.

But within the past century and a half, we humans have tipped the balance, probably with major long-term ecological conse-

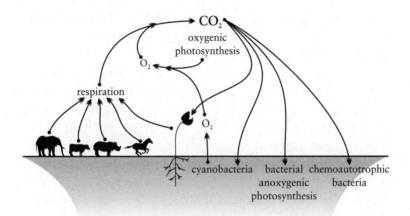

The biological carbon cycle.

quences. We have significantly augmented the atmospheric reservoir of carbon dioxide by burning massive quantities of carbon previously trapped in the form of coal or petroleum. And we have created our own special form of carbon trapping by manufacturing carbon-containing materials, principally plastics, that microbes, at least for the present, are unable to degrade to carbon dioxide. Although this accumulation can be unsightly and cause severe environmental problems, it is not nearly large enough to balance our use of fossil fuels. Nevertheless, the fate of plastics is not only a story that describes human perturbation of Earth's carbon cycle, but also an interesting tale of the potential usefulness of microbes.

Plastic Bottles on an Ocean Beach

Most beaches are littered with plastic detritus, bottles mainly, as are many other locations. In fact, it's difficult to walk on any beach and not encounter a few. If the beach has not been cleaned for some time, this plastic might be quite old and have floated there from great distances. Even remote beaches that are rarely visited and therefore not directly littered are indirectly littered, sometimes heavily so, by plastic that drifts to them.

Unfortunately, the piles of plastic we see strewn on beaches are only a trivial sampling of the vast quantities of plastic that are accumulating on our planet at ever-accelerating rates. One of the most massive repositories for such litter is the ocean itself. A particularly heavily littered region lies in the North Pacific Ocean in one of the areas called the *horse latitudes* because lack of prevailing winds there becalmed sailing ships, forcing crews to jettison their horses. It is also called the North Pacific Gyre because currents there flow in a circular (clockwise) direction. Floating debris becomes entrapped in this large whirlpool, earning for it more colorful terms: "Asian Trash Trail," "Trash Vortex," and "Eastern Garbage Patch." But "patch" is not quite an appropriate descriptor. The region is nearly the size of a continent. Mammoth quantities of floating plastic are entrapped there. Much of this debris, through the action of sunlight, mechanical abrasion, and wave action, has broken into smaller pieces. But the plastic remains. Some samplings have shown that in this region, the ocean contains six times as much plastic as phytoplankton. The cumulative impact on marine life, from the smallest creatures to birds and large mammals, is manifold. Some is simply mechanical. Birds and mammals swallow the larger chunks of plastic, which block their intestines. And there are toxic impacts as well on all sizes of marine life. Because many plastics act as chemical sponges, they concentrate chemical pollutants and become highly toxic, killing the organisms that consume them. Consuming plastics does not remove them. When the animal dies, it decomposes, but the plastic remains in the ocean. The Eastern Garbage Patch is certainly the largest such gyre in the ocean, but it is not the only one. About 40 percent of the world's oceans are classified as gyres. All are garbage collectors.

What is the basis of the special littering capacity of plastic? Lots of floating material—leaves, other parts of plants, animal remains

—enters the ocean. They may accumulate temporarily, but eventually they disappear because microbes consume them as nutrients. Microbes can consume almost any organic material.

Microbiologists have long subscribed to a fundamental dictum called *microbial infallibility,* which states that every naturally occurring organic compound can be degraded by some microbe or set of microbes. If an organism makes an organic compound, some microbe can degrade it. Evidence supporting the dictum is pretty good. Naturally occurring organic compounds do not persist indefinitely in nature because some microbe will use them as nutrients. Of course, some organic materials, for example peat and petroleum, persist for long periods. They are resistant to microbial decay, particularly in their natural anaerobic environments, but they can be broken down by microbes if oxygen is available. Plastics cannot. We might ask, why? For one reason, most plastics are highly insoluble in water, and nutrients enter most microbes, including prokaryotes and fungi, only if they are in solution. But that is not a complete answer. Cellulose, which is extremely insoluble in water, as insoluble as most plastics, is readily degraded by some microbes, such as those in the cow's rumen. The best explanation of plastics' resistance to microbial attack is that microbes evolved over the eons to degrade and utilize as nutrients the organic compounds that were present in their environment. They learned to use them as other organisms learned to make them. But plastics have only recently been added to the environment by humans. Some microbiologists argue that in time microbes will evolve to degrade plastics, but if they do, it will certainly be a long time. There are no signs of it yet.

The differential environmental impact of human-made substances that can and cannot be degraded by microbes is readily apparent. A dramatic example is provided by sulfonate detergents, which are highly effective for their intended use. When first introduced in the 1960s, they were composed of long, branching chains of carbon atoms with a sulfonate group ($-SO_4^-$), compounds that

were extremely resistant to microbial attack. The consequences of
their widespread use were disastrous. These detergents passed un-
degraded through sewage treatment plants and caused extensive,
unsightly, and environmentally damaging foaming to occur in bodies of water that received the outflow. Of course, the treatment
plants also foamed. The solution for this particular affront to the
dictum of microbial infallibility proved to be rather easy. Detergents with straight rather than branching chains of carbon atoms
can be degraded by certain microbes. When they replaced the
branched-chain detergents, the foaming stopped because microbes
in the sewage treatment plants degraded them.

Conversion to general use of plastics that microbes can degrade
would be a major environmental advance—imagine clean beaches,
a plastic bag–free view of the world, unthreatened ocean inhabitants. There are a number of candidate biodegradable plastics. All
involve processing natural materials instead of chemically converting petroleum to compounds such as plastics that never before existed on our planet.

One class of candidates that are already being produced is bacterial storage products, called *PHAs* (poly-beta-hydroxyalkonates),
the primary one being PHB (poly-beta-hydroxybutyrate). All creatures store, when conditions are good for them, reserves of excess food for subsequent use in leaner times. We, as we are all too
aware, store most of our overabundance of food as fat in special
fat-storing cells. We store a lesser amount of it for ready use in the
form of glycogen in our muscles and liver. Unlike us, when times
are good, bacteria do not accumulate fat. A few, including our
friend and sometimes antagonist *E. coli,* store their excess bounty
as glycogen. But many store it as PHAs for later use. And some bacteria under the right conditions—when lack of one nutrient pre-
cludes growth but their major organic nutrient is still available in
excess—store a lot of it. Under such conditions the gram-negative
bacterium *Alcaligenes eutrophus,* for example, will store over 80
percent of its dry weight as PHAs—a new concept of obesity. Un-

surprisingly, the bacteria that store excess nutrients this way degrade PHAs and use them as nutrients when needed. Many other microbes, as well, can break them down. PHAs do not accumulate in nature.

Remarkably, PHAs have properties quite similar to petroleum-based plastics. Just about any familiar plastic item—bottles, plastic sheets, bags—can be made from PHAs. The idea of using PHAs in place of petroleum-based plastics is not new—patents for such use were filed in the United States in 1962. But such PHA-based plastics were not made industrially until 1982, when Imperial Chemical Industries of Britain marketed them under the trade name of Biopol. The first consumer product, bottles for a biodegradable shampoo, was marketed in 1990 by the German corporation Wella AG.

The commercial process for making PHAs is particularly attractive because the source of its carbon is carbon dioxide. The organism, *Alcaligenes eutrophus,* derives its energy from an inorganic reaction, the oxidation of hydrogen gas. Only carbon dioxide, hydrogen, oxygen, and of course *A. eutrophus* are needed to make this biodegradable plastic. But process costs are such that it is significantly more expensive than petroleum-based plastics—though only if the costs of environmental impact are neglected. Still, with the prices of both petroleum and environmental cleanup escalating, microbial plastics almost certainly have a future. Perhaps one day no bottles will be seen on a remote beach. Earth's carbon cycle, though perhaps still out of balance owing to other human activities, would again be functional.

8

Hostile Environments

Many microbes thrive in places that we would consider extremely hostile, almost unbelievably so. Such environments strike us as being much more lethal than nurturing. But not to some microbes. Collectively the microbes that thrive in these environments, most of which are archaea, though some are bacteria, are called *extremophiles*. In this chapter we will encounter microbes that live in saturated solutions of salt, at temperatures hotter than boiling water, and in ocean depths with crushing hydrostatic pressures. We will even meet some that can grow vigorously well below the freezing point of water. We have already looked at the extremophilic microbes that live in Earth's most acidic environments, such as copper mines, and we will save those that survive witheringly intense radiation for our farewell sighting.

Red and Green Patches in San Francisco Bay

If you are sitting by a window on the right side of a plane approaching San Francisco from the east on a clear day, you will notice in the water on the east side of San Francisco Bay large red patches, some redder than others. Looking more closely, you will notice adjoining ponds of various shades of green, some pale, others richly verdant, resembling a pampered golf green. Many other such patches of red water exist throughout the world, but these from this vantage might well be the most spectacular.

What we see is a saltern (a salt-making facility). The patches are diked-off evaporation ponds in which table salt (sodium chloride) is being rendered from seawater. San Francisco Bay is a favored location for this because, in spite of its relatively cool climate, evaporation rates are particularly rapid, owing to low rainfall and steady breezes. For solar production of salt to be economically feasible, evaporation must exceed rainfall by about threefold. That is the case in San Francisco Bay as it is in other regions with Mediterranean climates and, of course, in desert regions as well. Humans have been making salt this way for a very long time. They have made it by evaporation near the Dead Sea in Israel and Jordan since biblical times. Mined salt, a market that expanded markedly in the nineteenth century, is the alternative to sea salt, although it is not fundamentally different. Underground salt deposits are the residuals of ancient seas. Nevertheless, some chefs express a passionate preference for sea salt. Certainly, sea salt sounds more romantically appealing than rock salt or mined salt.

An extraordinary, and to me incomprehensible, mystique has built up around salt preferences. Not only are claims made for the superiority of sea salt, they are made for salt from specific regions of the sea. For example, Maldon salt from the east coast of England, Kilauea salt from waters off Molokai, and Celtic sea salt from Brittany are particularly prized and they bring premium prices. Kilauea salt sells for about forty dollars a pound. Exotic salts are touted as being healthier, tastier, and naturally gray, or in one case, almost black. Kosher salt, another oft-cited preference of chefs and gourmet recipes, owes its distinction to the large size of its salt particles and lack of additives, notably iodine. "Kosher" does not in this instance refer to the method of manufacture of salt, but to its use in preparing kosher meat. Gourmet specialty stores frequently offer salt grinders, implying a taste or health advantage of freshly ground salt.

Most table salt in the western world is ground into smaller grains by the manufacturer, and iodine is added to it. This simple and inexpensive practice (iodized salt was offered at no increase in price), which began early in the twentieth century, has had a dramatic impact on public health. By eliminating iodine deficiency, goiter (swelling of the neck owing to enlargement of the thyroid gland), cretinism (a severe mental and physical defect resulting from iodine deficiency in infancy), and iodine-deficiency mental retardation have almost been totally eliminated from countries where iodized salt is used. Still, however, iodized salt is available to only about 70 percent of the world's households.

We probably consume too much salt for optimal health. The American Heart Association recommends that we restrict our daily consumption of sodium to three grams, which corresponds to about 7.5 grams or a teaspoon and a half of salt. But we cannot avoid it altogether. Sodium ion is an essential component of a human diet. Athletes who sicken and sometimes die from drinking excessive quantities of water do so largely because they have massively deranged their electrolyte balance, principally as a result of depleting their reserves of sodium ions. Salt has also been used since ancient times as a food preservative, because high concentrations of salt prevent the growth of microbes. Salt pork, kept in barrels of brine, was vital sustenance for sailors on long voyages as well as for frontier farm families during winter months. Meriwether Lewis and William Clark and the Corps of Discovery relied heavily on salt pork during their expedition at the beginning of the nineteenth century.

Obtaining salt, transporting it, and taxing it has played a sometimes determinative role in human history. We have hints of salt's pervasive impact by recalling that our word "salary" derives from Roman soldiers being paid in salt. Jesus called his disciples the "salt of the earth" to acknowledge their value to him. In the modern era, the British-imposed tax on salt was a catalyst for Indian independence with greater impact than the tea tax on the American colonies. In the seminal act of protest that began the movement of nonviolent civil disobedience, leading eventually to Indian independence and later to the U.S. civil rights movement, Mohandas Gandhi in 1930 defied the 1882 salt act that gave the British a monopoly on salt, its production, and their power to tax it. Gandhi, with a throng of supporters following, marched 240 miles from his village to the Arabian Sea, where he picked up a bit of salt and urged others to recover their own sea salt, without paying tax, of course, garnering worldwide attention. Within a month Gandhi was imprisoned along with many of his fellow protesters. The movement was under way.

Modern sea salt manufacture is hardly a romantic process, though it does provide an interesting view from an airplane. The method employed to manufacture salt from sea water establishes a set of massive culture vessels, a few feet deep and several acres in area, with different salt concentrations and nutrient content, each suitable for a particular kind of microbe. The variety of colors of the saltern ponds (salt evaporation ponds) that we see out the window of our airplane are those of the microbes that a particular pond contains.

Seawater is pumped into the first pond and allowed to flow by gravity into successive ponds. The concentration of salt in each successive pond is higher than in the previous one, and these concentrations remain approximately constant over time, as their colors do. It is a continuous process. The ponds are fed in series rather than in parallel because the saltern does more than just recover the mixture of substances dissolved in seawater, it purifies sodium chloride from seawater. San Francisco Bay water is only about half as salty as seawater, so the range of salt concentrations and microbe diversity among these saltern ponds is greater than those seen in the many salterns elsewhere fed by seawater.

Seawater contains about 3.5 percent dissolved material, approximately 77 percent of which is sodium chloride (NaCl). "Salt" is a generic term for ionizable substances, in spite of the fact that we commonly use it as a synonym for sodium chloride. The rest of the dissolved material, therefore, is chiefly salts of calcium (Ca^{2+}), magnesium (Mg^{2+}), and sulfate (SO_4^{2-}) ions. As evaporation proceeds, the first salt to precipitate out is gypsum (calcium sulfate). Then, when the salt concentration rises to about 25.8 percent, sodium chloride begins to crystallize at the bottom of that pond, leaving magnesium salts still dissolved. By using a sequential set of ponds, the sodium chloride is relatively pure when it crystallizes at the bottom of one particular pond. When about ten inches of salt have accumulated, it is harvested mechanically, washed, dried, ground, amended with iodine, and in some cases treated with an agent (calcium silicate) to make it flow easily. Some of you might remember the Morton Salt slogan, "When it rains, it pours," and the iconic umbrella girl in the rain, happily spilling salt from a leaking package.

The liquid that remains after harvesting the sodium chloride, containing the still-soluble magnesium salts along with some potassium chloride, is called *bittern* owing to its distinctly bitter taste. Bittern is used today to control dust on dirt roads. In ancient times, when the final step of salt manufacture was complete, it was important to remove the bittern before the sun completely dried the pond. The decision of when to do this was based on the color of the pond, i.e., relying on the microbes' ability to sense the concentration of sodium chloride.

The succession of microbes that develop in the various ponds of a seawater-fed saltern (the scientific term, far too elegant to pass up, is *thalassohaline*) and their colors provide a vivid display of the salt tolerances and preferences of halophilic (salt-loving) microbes. The first ponds in the saltern sequence—those with the lowest concentration of salt—are green because of the abundant presence of halophilic green algae. Then, in ponds with somewhat higher salt

concentrations, another green alga, *Dunaliella salina,* predominates. "Green alga" is a term that describes what kind of alga it is,
not necessarily its color. This single-celled alga that swims with the
aid of two flagella produces large amounts of beta carotene, which
gives the ponds it dominates a bright red color. These ponds with
intermediate salt concentrations are also home to the brine shrimp,
which offer some color of their own. The algae in such ponds contribute more than just beautification. They speed the rate of evaporation by increasing absorption of radiant energy.

Succeeding ponds become too salty for algae to tolerate. Halophilic bacteria thrive in these next ponds with higher concentrations of salt, and finally extremely halophilic archaea predominate
in those with the highest concentrations of salt. Thus representatives of the major groups of the entire microbial world are represented in the ponds of a saltern.

Archaea tolerate the highest levels of salt—perhaps unsurprisingly because they hold most biological records for surviving and
thriving in hostile environments, including those with blistering
temperatures and corrosive acidity. They are found in the crystallization ponds, which sometimes reach concentrations as high as
30 percent salt, an extremely hostile environment for most living
things. Among other deleterious impacts, high concentrations of
salt inactivate most proteins. Very high concentrations of salt stop
the growth of all microbes. That is why immersing food, such as
meat, in brine solutions is such an effective way of preserving them.
Bittern has too high a salt concentration and is too hostile an environment for any microbe. It is sterile.

The primary imperative for living in a high-salt environment is to
avoid intracellular desiccation as a consequence of water being
drawn out of cells by osmosis. The various microbes that live in
salty environments meet this challenge in different ways. Almost all
do so by increasing the concentration of small molecules inside
their cells in order to balance the concentration of salt on the outside. The small molecules they employ differ depending on the spe-

cies, but usually they are benign organic compounds (called *compatible solutes*), which do not damage the cell's proteins. But the highly salt-tolerant archaea that live in extremely salty environments such as crystallization ponds take a different approach. They use salts, largely potassium chloride, to balance the outside salt concentration.

Of course, this approach would be disastrous for most proteins, including a cell's vital enzymes. High salt content denatures proteins, that is, it changes their three-dimensional structure so that they lose biological activity and frequently become insoluble. The effect can be visualized by sprinkling a few grains of salt on uncooked egg white. Opaque white regions form around the grains as the proteins in the egg white become denatured and lose their solubility. Halophilic archaea avoid this problem by having evolved proteins that are intrinsically resistant to being denatured by salt.

The archaeon *Haloquadratum walsbyi* (named for the British microbiologist, Anthony Walsby, who first observed it), which proliferates in crystallization ponds, is unusual even among the halophilic archaea. That becomes clear just by looking at it under the microscope. Most of its cells are in the shape of perfect squares,

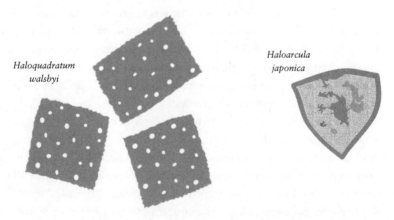

Haloquadratum
walsbyi

Haloarcula
japonica

Haloquadratum walsbyi and *Haloarcula japonica*.

coincidentally rather like salt crystals, with sharp corners and straight-line sides. Curiously, the colonies it produces when cultured on solid media are also square.

H. *walsbyi* is the only known square-shaped microbe. One of its close relatives, *Haloarcula japonica*, is rather unusual-looking too. It has triangular or rhomboid-shaped cells. Most halophilic archaea look like typical prokaryotes, with spherical or rod-shaped cells. *H. walsbyi*'s cells are also flat, and they appear to be peppered with small, whitish areas. These are gas-filled chambers (gas vesicles) that give the cell buoyancy and, thereby, access to air (oxygen is not very soluble in saturated brine). Such buoyancy also exposes *H. walsbyi* to sunlight, which, as we will see, is critically important to its lifestyle. And *H. walsbyi* can swim as well as float. Each cell produces one to several flagella, conferring the ability to move actively through its very salty environment.

H. walsbyi not only tolerates high concentrations of salt, it prefers a briny cauldron that contains 25 to 35 percent salt at a temperature between 42 and 52°C (108 and 126°F). Its red cells confer the spectacular red color we saw from the airplane for the same reason that the alga *Dunaliella salina* reddens the ponds with lesser concentrations of salt. *H. walsbyi* produces a carotenoid, a group of compounds that include beta carotene, which our bodies can convert to vitamin A. Carotenoids protect both *H. walsbyi* and *D. salina* from the destructive effects of intense sunlight.

A fascinating feature of *H. walsbyi* and other halophilic archaea is their ability to carry out a unique and extremely minimal form of photosynthesis. Unlike all other forms of photosynthesis, plant or microbial, this form of photosynthesis does not depend on the green pigment chlorophyll. And it does not produce oxygen, as many forms of photosynthesis do, it just produces metabolic energy—not very much, but adequate amounts to support a slow rate of growth and proliferation, as well as to swim a bit and to rid its interior of some of its salt.

This form of photosynthesis is the ultimate in simplicity. When oxygen becomes limiting or nutrient supplies are depleted, these microbes resort to it by making purple-colored patches in their cell membranes. Such purple membranes, as they are called, contain a lattice structure of a protein, bacteriorhodopsin, that is closely related to the protein rhodopsin, which is found in the retina of our eyes—the one that allows us to perceive light. Bacteriorhodopsin, like rhodopsin, is bound to a small molecule, retinal, which we synthesize from vitamin A (the basis for the traditional admonition to eat carrots for better night vision). The halophilic archaea do not need to synthesize it from vitamin A—they make it from scratch.

Bacteriorhodopsin spans the cell membrane, and retinal is attached somewhere in the middle. Retinal has chemical properties that allow it to pick up and release a proton (a hydrogen ion, H^+). It picks up this proton on the inside of the cell; then when a photon of light energy strikes the protein, it releases the proton outside the cell. Thus, the purple membrane acts as a patch of light-driven proton pumps, driving protons out of the cell and, thereby, forming a proton gradient: high outside, low inside. As I mentioned in Chapter 3, this gradient represents potential energy tending to force protons back into the cell through ATP synthase, thereby generating ATP and hence metabolic energy.

Bright Colors in the Hot Springs of Yellowstone National Park

The medley of colors in the outflows from hot springs in Yellowstone National Park is a gorgeous and complex package of microbe sightings. Yellowstone is not the only place where hot springs with such colorful displays can be seen, but it is certainly the most extensive and the most impressive. Volcanic regions of Iceland, Italy, Japan, and New Zealand for the most part lack the unspoiled richness of Yellowstone because they have undergone development as spas and as suppliers of geothermal power.

In certain respects, the present environment of Yellowstone mimics Earth's very hot environment 4 billion years ago when microbes were first evolving, and for similar reasons. Earth's crust was thin when the planet first formed. It is still thin in the region of Yellowstone, only about forty-five miles thick as compared with about ninety miles in most places. At certain locations in Yellowstone where there are upwellings of molten rock, it is thinner still. In some of them, molten rock is only two or three miles below the surface. But it is not necessary to go all the way down to the molten rock in order to encounter intense heat. When water seeping into the ground approaches these hot spots, it spews back as a geyser or a hot spring and then cools progressively as it flows away, exposed to Yellowstone's rather cool climate. So the outflow of a hot spring or geyser provides a continuum of temperature environments from near boiling to somewhat chilly, rather like a saltern provides a near continuum of salty environments. The particular microbes one finds in the outflow change with distance from the source. Noting water temperature, color of the microbial masses, and the ways that cells aggregate (in filaments, flocs, slimes, scums, or mats) to form these masses gives us a pretty good idea of which microbes are present.

The spectacular microbe-watching opportunities offered at Yellowstone are facilitated by the region's openness. Because hot springs occasionally overflow and flood larger areas, their surroundings are almost completely devoid of trees and most other plants. In response to the intense sunlight, many microbes in the springs produce colorful, protective sunblock pigments, mostly carotenoids. Most of these microbes that are so brightly colored in sunlight are colorless in the dark. Prudently, they use sunscreen only when needed. But not all of the colors in outflows from hot springs are microbial. Some are mineral precipitates; these mineral deposits, however, are not easily distinguished from microbes. They do form in the hotter regions as red iron and yellow sulfur deposits but only at the water's edge, not midstream where most of the microbes are.

A good place to begin our sampling of Yellowstone's array of microbe sightings might be a spring such as Octopus Spring near the Great Fountain Geyser in Lower Geyser Basin. Although access to this spring is now restricted owing to its environmental sensitivity, there are many other springs like it. Octopus Spring has special status for microbiologists, however, because it was here that *Thermus aquaticus,* sometimes called the "300 million dollar bacterium," was discovered by Thomas Brock during his studies of Yellowstone's thermophilic microbes in the 1960s. In the spring's outflow at the point where the temperature has cooled to 163°F (73°C), extensive orange-colored masses of cyanobacteria are readily apparent. Close by though less obvious, whitish, long stringy masses are clearly discernible. These filaments are *T. aquaticus.*

The proximity of the two microbes is not accidental. Cyanobacteria make their own food supply by photosynthesis from carbon dioxide, producing enough to satisfy as well the needs of *T. aquaticus,* which, unable to utilize light energy, must have an external source of organic nutrients. In this hot spring, cyanobacteria are that source. In other hot springs *T. aquaticus* grows more sparsely on its own at the expense of the minute amounts of organic nutrients that are present in the spring water as it bubbles from the ground.

T. aquaticus gained notoriety when it became a critical contributor to the polymerase chain reaction (PCR) method of detecting and dealing with minute amounts of DNA, which has exerted such a profound impact on modern life. PCR is applied to DNA-based criminal convictions and exonerations as well as medical diagnoses of diseases such as HIV-AIDS. PCR involves multiple copying of a small sample of DNA (that contained in a drop of blood, for example). Each round of copying doubles the amount of DNA in the sample. The copying enzyme used, the same one that cells use to copy their own genetic storehouse in order to pass it on to progeny, is DNA polymerase. Because DNA polymerase only copies single strands of DNA, naturally occurring DNA in its double helix (double-stranded) form has to be separated into single strands by heating (melting) it before it can be copied. So between each round of copying in the PCR procedure, the sample of DNA must be heated to near boiling, a treatment that would inactivate the DNA polymerases from most organisms. But not the one from *T. aquaticus*. Its polymerase, dubbed Taq (for *T. aquaticus*) polymerase, readily withstands such temperatures. If other polymerases were used, a fresh amount of Taq would have to be added before each

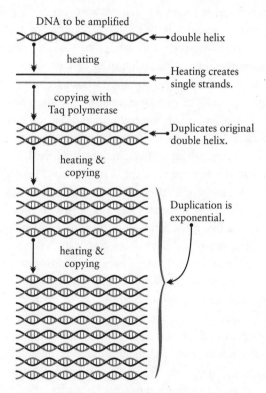

Explanatory sketch of PCR. Not shown in this sketch is the necessary addition of "primers," small bits of DNA that tell the polymerase enzyme where to start copying a single strand of DNA. Such primers focus amplification on a particular region of DNA, one of particular interest, and are required for copying owing to DNA polymerase's inability to copy without them.

round of copying. Taq polymerase made PCR easy and very widely used. Kary Mullis, then an employee of Cetus Corporation, had the idea for PCR and for using Taq polymerase. Later Roche Molecular Systems bought the PCR patents from Cetus for 300 million dollars, raising *T. aquaticus* to its special economic status. Taq polymerase was also a big winner: in 1989 *Science* magazine named it "Molecule of the Year." Mullis got only ten thousand dollars, but was consoled with a Nobel Prize. Neither the microbiologist Thomas Brock nor the microbe *T. aquaticus* received a penny.

Other notable denizens of the near 80°C regions of outflows from hot springs in Yellowstone are species of *Chloroflexis*, although none are moneymakers. This bacterium can be recognized by its formation of beautiful beige to orange to orange-red mats near or on top of green mats of cyanobacteria, providing a striking color contrast. *Chloroflexis* is photosynthetic and, as molecular studies show, it is quite an ancient microbe, having flourished in this sort of environment since shortly after Earth formed. Its bizarre form of photosynthesis might have been the first to evolve. Like all forms of photosynthesis, the one in *Chloroflexis* uses carbon dioxide as a source of carbon atoms (for its PMs and hence all its cellular constituents). But unlike most forms of photosynthesis (including those of green plants, algae, and cyanobacteria), it does not produce oxygen gas as a byproduct. That is because *Chloroflexis* and other members of this small microbial group (inelegantly but descriptively termed *green nonsulfur bacteria,* now properly but rarely called *chloroflexi*) do not use the hydrogen atoms from water to reduce carbon dioxide to make their PMs. Instead, most chloroflexi use hydrogen atoms from organic compounds, although some can use hydrogen gas (H_2) directly. These two forms of photosynthesis can be contrasted as follows: photosynthesis by plants, algae, and cyanobacteria begins with light energy, carbon dioxide, and water, and produces PMs and oxygen; whereas photosynthesis by chloroflexi substitutes hydrogen atoms for water, and so produces no oxygen.

Chloroflexis can also ferment organic compounds in order to obtain metabolic energy. Thus it seems ideally suited to primitive Earth with its high temperatures, lack of oxygen (plants, algae, and cyanobacteria had not yet evolved), and relative abundance of chemically formed organic materials. Those ancient fossils that have been presumed to be filamentous cyanobacteria might very well be chloroflexi. They, like cyanobacteria, are filamentous, and certainly they are more suited to Earth's early environment.

Because of its eerie beauty, one of the most interesting microbe sightings in Yellowstone is the delicate pink filaments that wave pennantlike in the outflow of Octopus and other springs at a temperature of about 90°C (194°F). Thomas Brock first mentioned it the 1960s. Brock was able to cultivate the 300-million-dollar bacterium *Thermus aquaticus,* but not the pink filament-producing microbe. Although *T. aquaticus* itself is pink, it is not the filament former, which is secured to a solid surface. *Thermus aquaticus* is not swept away by the flow and is present only because small numbers of its cells are stuck to the actual filament former. Numerous attempts to cultivate the primary filament former failed. That is not too surprising because the overwhelming majority of microbes one sees in nature defy attempts to cultivate them in the laboratory. Molecular studies on the filament former's DNA showed that it is a bacterium. That was surprising because archaea dominate in such high-temperature environments. It is related to a primitive group of hyperthermophilic bacteria (the Aquificales) that derive enough energy by respiring hydrogen gas to make their PMs and cellular constituents from carbon dioxide.

In 1998, a German microbiologist, Karl Stetter, was successful where many others had failed. He was able to cultivate the filament former using the meticulous approach of devising a culture medium based on actual chemical analyses of water in Octopus Spring. He named the bacterium that he cultivated *Thermocrinis ruber.* It grew well in the medium he devised in which it respired a hydrogen sulfide (H_2S) to generate ATP and used carbon dioxide as a source of its PMs. But disappointingly, in culture it grew as individual cells, not as filaments. Stetter then mimicked the natural environment a little more precisely. He cultivated *T. ruber* in a flowing stream of nutrients and, indeed, then the cells did aggregate into delicate pink filaments, just like those Thomas Brock marveled at in the 1960s.

So when you visit Octopus Spring or similar springs in Yellowstone, you can be assured that the pink filaments in the outflow are aggregates of *Thermocrinis ruber*, a very primitive, hyperthermophilic bacterium, and they are probably respiring hydrogen sulfide to obtain energy and using carbon dioxide to make their cellular constituents. You can smell the excess hydrogen sulfide that the microbes did not use.

You might find it odd that we have not discussed archaea in Yellowstone, because archaea are the record-setting hyperthermophiles and Yellowstone is an extensive hyperthermophilic environment. They are certainly present. Molecular studies, which isolate and sequence DNA instead of attempting to cultivate living microbes, have established the abundant presence of diverse populations of archaea in Yellowstone's hot springs. They grow for the main part as individual cells, too small for us to see. They are also extremely difficult to cultivate. But there is an astounding variety of archaea in Yellowstone's hot springs. One molecular study found seventeen different species of archaea, seven of them previously unknown, in a single hot spring (Jim's Black Pool).

Yellowstone displays an abundant variety of plants, animals, and natural beauty, but its largely ignored microbial diversity, and perhaps uniqueness, might be the most spectacular of its treasures. And the microbes were there first.

Moist Greenish Patches Near a Hot Spring

Before leaving Octopus Spring in Yellowstone National Park, you should make at least one more microbe sighting. Take a careful look at the moist greenish patches nearby, in the areas where the pond overflows from time to time. They are masses of microbes, called *microbial mats*. Some are several inches thick; others are much thinner, some as slim as an eighth of an inch or less. Today, well-developed, thick microbial mats are seen only in geothermal regions and other hostile environments such as very salty marshes. Once microbial mats were widespread on our planet. They flourished before plants and animals intruded on the world to which microbes then had exclusive rights. But now animals eat them and plants shade their sunlight. Well-developed microbial mats are pretty much restricted to environments that are hostile enough to exclude these latter-day mat-destroying intruders. Microbial mats are not particularly spectacular or eye-catching, but they do tell us important tales about how microbes interact with one another as well as about the history of planet Earth.

Microbial mats are particularly complex communities of interacting microbes, but they are not the only ones. Interactions and interdependence between and among microbes are not unusual in nature. In fact, they are the rule. Less extensive, simpler interacting layers of microbes are called *biofilms*. They occur everywhere, forming among many other such microbial communities—the scum on our unbrushed teeth, the mucus in the lungs of persons suffering from cystic fibrosis, and the slippery slime on rocks in pristine streams. Microbes do not just stick together and coexist in these communities. They stimulate and nourish one another, communicating in sometimes seemingly sinister ways by sending and receiving chemical signals. For example, some microbes, as a consequence of growing in biofilms, are more resistant to antibiotics than they are when growing alone. Their resistance derives from more than being physically protected from the antibiotic because they are imbedded in the biofilm. Each cell changes to become intrinsically more antibiotic resistant. And some microbes, as a consequence of growing in a biofilm with another microbe, are capable of causing more severe illness than if they stood alone. As a case in point, a species of the bacterium *Burkholderi* becomes more virulent when growing with the bacterium *Pseudomonas aeruginosa* in a biofilm within the lungs of a person suffering from cystic fibrosis than when growing there alone.

A biofilm is not just a consequence of microbial cells sticking together after a random, casual encounter. Individual bacterial cells actively seek out other individual cells. This matchmaking tendency, perhaps compulsion, was illustrated in an intriguing way by some experiments conducted at Princeton University with microbiology's well-studied experimental workhorse, *Escherichia coli*. The experimenters added a somewhat diluted suspension of individual *E. coli* cells to a microscopic maze that they had constructed. At first the cells were evenly distributed throughout the maze. But within a few hours they had all congregated in a mass at a single location within the maze. The experimenters discovered that the cells had found one another by mutual attraction. Each cell entices other cells to come to it by secreting small amounts of the amino acid glycine. Other cells sense the presence of glycine in their environment and swim toward its source and highest concentration—a well-known microbial behavior called *chemotaxis*. Collectively, accumulating groups of cells produce more glycine than individual cells and act as more powerful attractants. Eventually, all cells in the maze accumulate in a single spot. There is no good explanation for why they excrete glycine. Any amino acid to which cells are attracted presumably would work equally well. Similar mechanisms of mutual attraction must be at the root of the formation of biofilms. And living together in biofilms or mats must offer microbes profound selective advantages in a hostile world.

The distinction between biofilms and microbial mats is not a sharp one. It is a quantitative not a qualitative distinction. But the several-inch-thick masses of microbes near Yellowstone's hot springs clearly qualify as microbial mats. These masses are trophic communities, food chains, if you like, that resemble the food chains of plants and animals. The mats are more metabolically complex, however, owing to microbes' greater repertoire of energy-yielding mechanisms.

An examination of which microbes are in the mat, how they acquire metabolic energy, and how they interact summarizes most of what we have gleaned from all previous sightings. Cyanobacteria are at the top of the mat and the bottom of its food chain. They are the community's most abundant and prolific primary producers. By their form of photosynthesis, called *oxygenic* (oxygen-producing) *photosynthesis*, they harness the energy from sunlight to acquire their metabolic energy. They make their PMs and hence cellular constituents from carbon dioxide, and they produce oxygen as a byproduct. As you will recall, oxygen is a byproduct of using the hydrogen atoms from water to reduce carbon dioxide in order to make PMs. We are familiar with this oxygenic form of photosynthesis because it is the same one that all plants and algae use. And as we will learn from future sightings, this metabolic similarity is not exactly a coincidence.

Eukaryotic photosynthesizers did not just copy cyanobacteria;

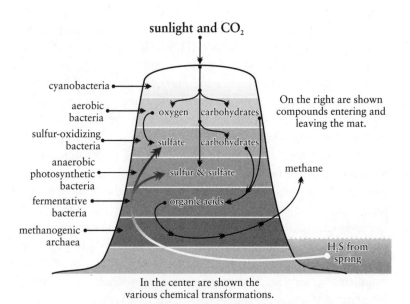

Mutual feeding in a microbial mat.

they used them. In the distant evolutionary past, a primitive eukaryote destined to be the progenitor of algae and plants captured
a cyanobacterium, making it an intracellular photosynthesizing
slave. The cyanobacterium's progeny were passed along with those
of the eukaryotic cell. Over the eons of their intracellular existence,
the captive cyanobacteria evolved and changed profoundly. But
even today their still-intracellular progeny, now called *chloroplasts,*
exhibit unmistakable metabolic and genetic evidence of their origin
as a free-living cyanobacterium. For example, they still contain
remnants of a cyanobacterium-like chromosome. Only cyanobacteria invented oxygenic photosynthesis. All other extant oxygenic
phototrophs use the intracellularly evolved cyanobacteria, chloroplasts, for that purpose.

Below the cyanobacterial layer at the mat's surface, in a region
made anaerobic by bacterial respiration, other bacteria carry out
anoxygenic (not oxygen-producing) photosynthesis. Some of these
phototrophs (the nonsulfur green bacteria that we sighted in the
hot spring) use organic compounds as a source of hydrogen and
others (green and purple sulfur bacteria) use reduced sulfur compounds such as hydrogen sulfide (H_2S). The hydrogen sulfide is
supplied directly from the hot spring. Most hot springs are sulfurous. Fire and brimstone go together, after all. The filtered light, the
light that the cyanobacteria did not absorb, that reaches this lower
level is ideally suited to the light-absorbing pigments of these anoxygenic phototrophs.

Other microbes, still lower in the mat, make their livings from
the offerings of the mixture of phototrophs above them. It is a complex, interconnected dependence, a food web rather than a simple
food chain. Higher in the mat, aerobic bacteria live directly off
the products—carbohydrates and oxygen—that the cyanobacteria
make. They use the oxygen to respire the carbohydrates, thereby
removing all oxygen and rendering the lower regions of the mat
completely anaerobic. In this lower oxygen-free zone, the anaerobic photosynthetic bacteria flourish and other bacteria ferment re-

sidual carbohydrates that the aerobic bacteria missed, producing organic acids as endproducts. These are converted by archaea into methane gas, which escapes from the mat.

The products of the anoxygenic phototrophs feed a separate but interconnected food web. Some bacteria (species of the genus *Beggiatoa*) in the upper regions of the mat generate their metabolic energy by oxidizing both the elemental sulfur (S^0) produced by the sulfur-using phototrophs as well as hydrogen sulfide directly from the spring to sulfate (SO_4^{2-}) using oxygen produced by the cyanobacteria. And other bacteria (sulfate-reducing bacteria, the ones that make the Black Sea and marine mudflats black) in anaerobic regions of the mat make their living by doing the opposite—reducing the sulfate that the sulfur-oxidizing bacteria produce at the expense of oxidizing some the organic acids produced by the fermenting bacteria.

It is easy to see how microbes benefit from living together. They feed as well as protect one another. Almost all the general forms of microbial metabolism are represented within the few inches of the microbial mat near Octopus Spring in Yellowstone as well as within microbial mats in other hostile environments.

Of course, the food web we just walked through functions only during the daytime, when the mat is bathed in sunlight. At night photosynthesis by the cyanobacteria stops, and everything changes. Suddenly, the mat becomes deprived of oxygen, and all downstream members of the food web must adjust. *Beggiatoa* takes an active approach. Being capable of an odd type of movement called gliding motility, it moves from the mat's oxygen-depleted interior to the mat's edge or upper surface, where it has access to atmospheric oxygen. Then in the morning, when sunlight turns on oxygen formation again, it moves back inside, where concentrations of its other required nutrient, hydrogen sulfide, are higher, another example of the benefits of microbial chemotaxis. Microbes are willing and able to move in order to improve their lives.

As we noted earlier, microbial mats, some quite large, were probably plentiful during Earth's early history when microbes were its exclusive inhabitants. Many of these ancient mats have been preserved as fossils known as stromatolites. They are the only fossils that record the first near 90 percent of the history of life on Earth. They constitute the sole remaining fossil record of Earth's earliest known ecosystems. Clear images of microbial cells that look quite like modern-day cyanobacteria have been obtained by examining thin slices of stromatolites under an electron microscope. Some of these are from ancient, flintlike rock formations called *cherts* in Western Australia, which are 3.5 billion years old. Without doubt, these fossils prove that microbes have existed on Earth for that long. But whether they are cyanobacteria is less certain—these microbial cells may actually be chloroflexi. They certainly look like cyanobacteria, but that raises other questions. Modern-day cyanobacteria produce oxygen. Indeed, they, together with their chloroplast slaves, are responsible for all of the oxygen in our atmosphere. But powerful evidence shows that our 3.5 billion-year-old atmosphere contained no or only undetectable levels of oxygen. If these ancient fossils actually are the remains of cyanobacteria, why did they have so little impact on Earth's atmosphere? That remains largely an open question.

Microbial mats in hot springs may appear to be simple, but they have a complex metabolic and evolutionary story to tell. Extremely hot environments are not the only unlikely places for complicated forms of life to thrive. We will turn now to microbes that thrive under crushing hydrostatic pressure.

Deep Sea Rifts and Hydrothermal Vents

Transportation to this microbe sighting is a bit hard to come by because it lies at such an extremely remote place—the bottom of the ocean near its middle, over a mile and a quarter down. And once you arrive, it is not very hospitable: it is not only pitch dark and cold, only a few degrees above freezing, but at 7,000 feet under the surface the hydrostatic pressure is crushing—more than 3,000 pounds per square inch. The only possible way to get there is in a pressure-resistant, battery-powered submersible vehicle such as Alvin, the research vessel that belongs to the Woods Hole Oceanographic Institution in Woods Hole, Massachusetts.

Such inhospitable environments dot both the Pacific and Atlantic oceans, but they were beyond our reach until 1977, when it became technologically feasible to journey there. Those who have visited deep sea rifts and hydrothermal vents are startled by their unique, stunning beauty. They are bountiful biological oases on the otherwise almost sterile ocean floor. Nothing like them, or some of their inhabitants, occurs anywhere else on the planet. Life at such a depth is quintessentially microbial. Darkness prevents photosynthesis, so all life there depends on bacteria rather than photosynthetic organisms, which supply nutrients to the rest of the planet. Here bacteria use carbon dioxide and respire inorganic nutrients, members of a class called *chemoautotrophs,* not too unlike those that live in mine drainage. They are exclusive, primary producers of metabolic energy, which they obtain by oxidizing hydrogen sulfide while using carbon dioxide as a source of carbon for their PMs. All the other organisms in this richly diverse ecosystem are heterotrophs, organisms that obtain both metabolic energy and PMs from organic compounds that the resident chemoautotrophic bacteria provide.

Such an ecosystem demands more than a sterile ocean floor, of course. The underwater geysers the microbes call home are clustered at mid-ocean. A line of such geysers runs around the world where Earth's intensely hot magma continuously wells up, forming a new outer layer of the sixty-mile-thick lithosphere. This layer spreads out toward the continents, where it dives under (subducts) the continental plates in deep-sea trenches (some five to seven miles deep) near the shore. The rich microbial community and other life dependent on it is found near where the magma meets the ocean water, near the hydrothermal vent.

The geysers form where water seeps down toward the magma, becomes extremely hot, and rushes back to the surface, still at a temperature of 750°F (400°C)—hundreds of degrees above the boiling point. Under normal conditions on Earth's surface, water boils at 212°F (100°C), and evaporating water in the form of steam prevents the temperature from rising further. But boiling does not occur and steam does not form so deep in the ocean because its intense hydrostatic pressure prevents boiling. So hydrothermal rifts offer two astoundingly extreme environments: intense hydrostatic pressure as well as intense heat near extreme cold.

There are two kinds of underwater geysers: black smokers and white smokers. Black smokers are the hotter of the two. They spew a mixture of iron and sulfide that forms a cloudlike precipitate of black iron sulfide, conferring the namesake color. White smokers are a bit cooler. They spew a mixture of barium, calcium, and sili-

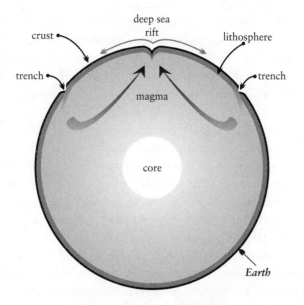

Mid-ocean spreading and trenches.

con that forms a white precipitate. On some of the smokers, such precipitates coalesce around the geysers to form chimneys that grow with astounding rapidity. One grew thirty feet during the eighteen-month interval between successive Alvin visits. Some chimneys attain the height of a fifteen-story building before they topple. Then new ones start to grow.

The biological communities that develop around these geysers are remarkably diverse and extremely unusual. Already, over 300 new species have been discovered living in them, all dependent on chemoautotrophic bacteria. The bacteria, like the ones growing in the outflow of some mines, obtain metabolic energy for themselves and the entire community by oxidizing the hydrogen sulfide that spews constantly from the vent. Hydrogen sulfide (H_2S) and oxygen (O_2) form sulfuric acid (H_2SO), releasing metabolic energy. They use that energy to fix carbon dioxide and thereby to supply the community with food in the form of organic nutrients.

It is sometimes said that this ecosystem is completely independent of photosynthesis. But that is not strictly true. It does need a supply of oxygen (O_2) for the bacteria to respire the hydrogen sulfide. And oxygen does not come from the vent. It diffuses down from the ocean's surface, where it is supplied from oxygen in Earth's atmosphere, all of which came from oxygenic photosynthesis, cyanobacterially based, as you will recall. But the site is spatially independent of photosynthesis. None takes place near the vents.

The chemoautotrophic bacteria are the primary producers of the vent ecosystem, the base of its food chain. They grow in thick mats near the vents, and amphipods (small, shrimplike crustaceans) and copepods (teardrop-shaped crustaceans with large antennae) feed on them. In turn, snails, shrimps, and crabs, including the elegant, endemic spider crab, feed on them. Several species of fish intervene at various levels of the chain.

Two animals in the community—the giant tube worm, *Riftia pachyptila*, and the giant hydrothermal vent clam, *Calyptogena*—opt out of the larger vent food chain and make their own arrange-

ments. The same hydrogen sulfide–oxidizing bacteria that form
mats near the vents also live inside some of the cells of these curious
animals, feeding them directly. The spectacular and beautiful giant
tube worm, which has become the poster child of the sea rift ecosystem, can grow vertically to be over eight feet tall and nearly an
inch in diameter. Its soft body tissues are protected by a nearly
snow-white tube made of chitin (the same substance the encases the
exoskeleton of arthropods such as crabs). The bottom of the tube is
firmly attached to rock near the vent. The top is decorated with a
brilliant red plume that waves from the tube. It is red because it is
drenched with a hemoglobin similar to the pigment in our blood
and in the nodules on the roots of leguminous plants discussed in
Chapter 5. The worm's hemoglobin is remarkable in that it exchanges three gases, H_2S, CO_2, and O_2, with the environment. Ours
in our red blood cells exchange only two, CO_2 and O_2; and the hemoglobin in root nodules just absorbs O_2. The red and white worm
has no digestive system. Instead, its residual gut region is packed
with green-brown spongy tissue called a *trophosome* (feeding
body). The bacteria, which make up about half the worm's weight,
live within the specialized cells of the trophosome. Thus, worm and
bacterium form a mutualistic symbiosis. The worm collects, concentrates, and supplies the bacterial cells with the three gases they
need, and in return, the bacterium supplies the worm with the organic nutrients it needs.

The giant hydrothermal vent clam *Calyptogena* is a variation on
this theme. It houses the cells of the hydrogen sulfide–oxidizing
bacterium within the cells of its gills. Like the worm's plume, the
clam's gills collect the gases the bacterium needs, and the bacterium
feeds the clam with the nutrients it needs.

All of the organisms that live near deep-sea rifts are extremophiles because they withstand, even flourish at, the high ambient
hydrostatic pressure. They are not crushed because the pressure inside and outside their cells is equalized as water passes readily in
and out of cell membranes. The limit of pressure a microbe can

tolerate is not set by what the cell can withstand. Instead, it is set by the pressure tolerance of vital chemical reactions occurring in the cell. Pressure resists any increase in volume, even those of molecules. And the volume of molecules usually changes in the course of a chemical reaction.

Before any reaction can proceed to completion, the participating molecules must pass through a transitional or *activated* state. If the molecular volume of this activated state is greater than that of the reactants, that is, if the reactants must expand to attain the activated state, high pressure will resist its formation as it does any expansion, thus slowing or stopping the reaction. Conversely, if the volume of the activated state is less than that of the reactants, pressure will favor its formation, thus speeding or enabling the reaction. A cell's most pressure-sensitive or pressure-dependent vital reaction, as set by the volume of its activated state, determines the maximum hydrostatic pressure that it can tolerate or benefit from. Some reactions are stopped by high pressure. Others proceed more rapidly at elevated pressures.

Natural selection being as unrelenting as it is, we should not be surprised that the activated states of some reactions in organisms inhabiting environments with high hydrostatic pressure are small. The bottoms of oceans are such an environment. The organisms that live there are called *piezophiles* ("pressure lovers"). Some microbes living in the deep ocean actually grow better at intense hydrostatic pressures. They are called *hyperpiezophiles*. One such bacterium, which is closely related to *Carnobacterium pleistocenium* and was discovered in a deep sea trench, prefers to grow at pressures 150 times higher than atmospheric pressure and can grow at pressures 600 times as high (8,800 pounds per square inch). Pressures that these bacteria prefer completely stop the growth of most microbes, and kill some. Indeed, the lethal effects of high pressures are now being used here on the surface to preserve foods, such as orange juice and guacamole.

Pressure is not the only extreme near a hydrothermal vent, of course. Microbes in these environments tolerate temperatures not found here on the surface. The record-setting heat-tolerant microbe *Pyrolobus fumarii* ("fire lobe of the chimney") was found here in the chimney of a black smoker. *P. fumarii* is an archeon that can grow at the astoundingly high temperature of 235°F (113°C)—twenty-three degrees higher than the boiling point of water at sea level.

In an environment of such microbial record breakers, we might also expect to find microbes that tolerate low temperatures in the nearby cold ocean water. Indeed, the record holder for growth at an extreme low temperature may also be found in an ocean, albeit far from a hydrothermal vent. Instead it lives in the polar ice at Point Barrow, Alaska. Polar ice is the ideal venue for extremely cold-tolerant microbes. Just as high pressure maintains liquid water at temperatures higher than the boiling point of water, polar ice is riddled with channels filled with liquid water. The water remains liquid below the freezing point of pure water because it contains high concentrations of salt. As polar ice forms from slush called *frazil ice,* the concentration of salt in unfrozen channels increases. The record-holding microbe is a bacterium, *Psychromonas ingrahamii,* named in my honor for studies I had conducted on bacteria that grow at low temperatures. Friends have reminded me that it does not move, is filled with gas, and is metabolically lethargic. But it is a record setter. It can grow at 10.4°F (−12°C), almost twenty degrees below the freezing point, doubling its numbers every ten days. *P. ingrahamii* is restricted to its cold environment. It cannot grow at temperatures higher than 50°F (10°C). Of course, it can also tolerate high concentrations of salt, up to 20 percent. Its environment must be salty enough to provide it with liquid water. The genus *Psychromonas* has adapted to the ocean's environmental challenges. Most species in the genus are psychrophiles, able to grow at low temperatures, and halophiles, able to tolerate elevated concentrations of salt. Some are piezophiles, benefiting from elevated hydrostatic pressure.

In addition to high and low temperatures and high pressure, extremely dry environments and desiccation are also challenges to all life forms, but microbes manage them amazingly well, as we will see.

A Package of Active Dry Yeast

Baker's yeast, available in any grocery store as a completely dry, beige-colored powder in a foil-backed package, is an unexpectedly remarkable microbe sighting. Even after prolonged storage, the yeast cells in the package burst into life when placed in warm water for a few minutes, and they become active fermenters of sugar, capable of converting it into alcohol and carbon dioxide. When yeast is used for baking, the carbon dioxide produced by the fermenting yeast cells becomes entrapped in the bread dough, causing it to rise. And the yeast gives bread that wonderful odor, completely different from the odor of bakery products raised with baking powder.

The successful production of active dry yeast that could be stored without refrigeration was a long-standing challenge met only during World War II, first by German and then by American producers of baker's yeast. Until then, baker's yeast was available only as moist cakes of packed yeast cells with a limited shelf life, even when refrigerated. Previous attempts to dry yeast invariably yielded cells that were dead and unable to ferment. The solution to the problem proved amazingly simple. It is only necessary to allow the yeast culture to stop growing by letting it exhaust some nutrient other than sugar and then holding it for an appropriate period of time in a nongrowing state before separating out the cells and drying them. When this procedure is followed, yeast cells remain active even after having been completely desiccated for years.

The basis for yeast's survival is its formation of a simple sugar, trehalose, when it is in a nongrowing state. Trehalose is chemically similar to sucrose or lactose in that they are all composed of two linked simpler sugars. For this reason they are called *disaccharides,* "two sugars." Sucrose is composed of a molecule of glucose linked to a molecule of fructose. Lactose is composed of glucose and galactose. Trehalose is composed of two molecules of glucose joined in a particular way, called an *alpha linkage.* It is the trehalose that yeast cells make that protects them as they are being dried.

The cell membrane is the vital structure that is particularly vulnerable to being torn apart and destroyed by drying. Cells with ruptured membranes are dead. Trehalose fits precisely between molecules in the membrane, holding them together so they do not fly apart when subjected to the disrupting stress of drying. Other disaccharides with similar structures, including sucrose, also offer some protection, but trehalose is the most effective by far.

Microbes are not the only organisms that make trehalose to survive desiccation. Small animals called "water bears" or tardigrades might be the most dramatic examples of this ability. These tiny animals, only about 1.5 mm long, make large amounts of trehalose and are able to come back to life after having been completely desiccated for years. When a piece of dried moss that had been kept in a museum for 120 years was moistened, active rehydrated water bears walked out of it as chipper as they must have looked many years before when the moss was dried.

In their desiccated state, water bears are highly resistant to other environmental assaults, including temperatures as low as −200°C (−328°F) and as high as 151°C (303°F). They can also survive intense radiation. But in spite of a water bear's remarkable abilities to survive, microbes, without question, hold all records.

It is common practice in microbiology laboratories to preserve stock cultures of microbes by keeping them at low temperatures, usually −20°C or less. Cultures are known to remain viable for years under such conditions. Still, a 2007 report by K. D. Bidle, S. Lee, D. R. Marchant, and P. G. Falkowski in the *Proceedings of the National Academy of Sciences* that bacteria had survived 8 million years in a frozen state was a distinct surprise. These microbiologists sampled ice from the dry valleys of Antarctica—the oldest-known ice on Earth—and recovered viable bacteria from them. The bacteria clearly showed signs of the ravages of age—particularly in their DNA, which is estimated to have a half-life of 1.1 million years. The damage to their DNA came largely from cosmic radiation. The age-weakened bacteria divided very slowly, a division taking almost six months in a rich medium. Nevertheless, they were alive after millions of years. It appears from these studies that 8 million years is near the maximum that bacteria and other microbes could survive in a frozen state, making it unlikely that microbes could have come to our planet on the icy head of comets from beyond our solar system, a trip that would have taken much longer than 8 million years. But microbes certainly would have had time to come this way from Mars.

The world's most resistant and persistent biological entities are bacterial endospores, which we will visit again in a later chapter. These structures can survive for hundreds, possibly thousands of years under ambient conditions. Dried, wrapped plant materials stored in botanical museums for hundreds of years have been found to contain viable endospores, capable of germinating to produce vigorous bacterial cultures. Most reports of even hardier endospores are suspect because of the possibility that spores in the original material germinated sometime during storage to produce a new crop of younger endospores, or because the sample may have been contaminated with endospores from another source sometime during storage.

Possibly the strongest evidence for endospores surviving thousands of years comes from the exploration of a flooded Roman fort, Vindolanda, dated 90 to 95 A.D. Viable endospores of *Thermoactinomyces vulgaris* were recovered there. *T. vulgaris* requires oxygen and a relatively high temperature to grow. Neither of these requirements was met by the fort's cold environment, made anaerobic by a rising water table. Because the fort was sealed it seems unlikely that new endospores entered it after a new sealing floor was added. So probably the endospores recovered from Vindolanda did, indeed, survive there for almost 2,000 years.

Often, we are more interested in killing microbes than appreciating how long they can survive. We do so for many reasons, including preserving food by canning or not contaminating certain environments—other planets, for example, by sending spaceships there. So the rate at which microbes die has been the focus of careful study. The results of these studies are unexpected. The rate of death of microbes when exposed to a lethal condition is different from that of other creatures. If a population of ants, for example, were exposed to a lethal temperature, the rate of their death would follow that of the well-known bell-shaped curve: after a brief exposure, those few most vulnerable ants would die; then as exposure

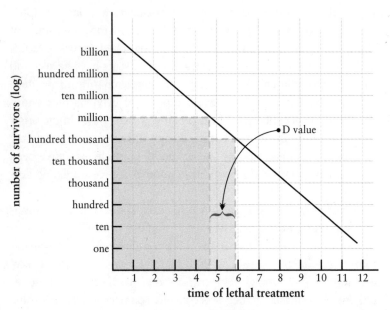

Rates of microbial death.

was continued, the great mass of ordinarily resistant or tough ants would succumb; and finally, after a more prolonged exposure, those few most highly resistant ants would be eliminated.

When exposed to a lethal condition, for example, an elevated temperature, microbes die with quite a different pattern. After a certain interval of time, 90 percent of the population will be dead: the population is decreased tenfold. Then, during succeeding intervals (called the *decimal reduction time* or D value for that particular microbe at that temperature), the population of survivors also will be reduced tenfold, and so on through succeeding decimal reduction times. Microbes are killed by a lethal treatment not from cumulative exposure as one might expect, but as though there were an equal probability of being killed during each time interval of exposure. A helpful if unpleasant analogy might be shots that were fired randomly at a constant rate into a crowd of people. The number still standing would decrease by the same factor over each equal interval of time because the number of targets continues to decrease. So to determine the treatment needed to sterilize (eliminate all microbes) from a particular environment, then, we have to know the size of the microbial population present and its D value.

The pattern of a microbial death curve has another curious aspect: it does not go to zero; it projects to fractions of microbes instead. This introduces an element of probability for predicting when no viable microbes will remain in the treated sample. So when considering a sterilizing treatment, one must also decide on the desired degree of assurance that the treatment will sterilize.

When the U.S. space program planned to send the *Viking* spacecraft to Mars, there was general agreement that its nose cone should be sterile so as not to contaminate the red planet with Earth's microbes. Heat was selected as the sterilizing agent, and an international group of microbiologists met to decide on the likely size of the microbial population on the nose cone, its D value for the planned treatment, and desired assurance of achieving sterility through heating.

Even an object as mundane as a package of active dry yeast can show us that microbes can persist for extended periods in a resting state. Neither heat nor cold nor dessication can eliminate them completely from a particular environment. Such harsh treatments can only strongly increase the odds.

9

Fungi, Hostile and Benign

Most fungi are eukaryotic microbes with tubelike cells called mycelia that are not visible without magnification. A few, called *yeasts*, are single cells or small groups of cells. All other eukaryotic microbes are protists. But sometimes masses of fungal mycelia aggregate to form visible structures called *mushrooms*. Although the vast majority of fungi play vital roles in nature, decomposing principally plant remains, some cause devastating diseases of plants, and others attack animals, including us, often by producing toxins. The soil is full of fungi. It is estimated that one gram of typical soil contains 30 to 300 feet of fungal mycelia. In this chapter we will look at some representative fungi.

A Couple of Sixteenth-Century Paintings

We have to depend on two world-famous sixteenth-century paintings, one by Matthias Grünewald and the other by Hieronymus Bosch, to remind us of a microbe that no longer troubles most of us: *Claviceps purpurea*, the fungus that causes ergotism, which had a devastating impact on medieval Europe.

The paintings themselves are arresting. Grünewald's painting, regarded by some scholars as an artistic accomplishment rivaling such Western cultural icons as Leonardo da Vinci's *Mona Lisa* and Michelangelo's ceiling in the Sistine Chapel, is a crucifixion scene now at the Musée d'Unterlinden in Colmar, France. It was painted as an altarpiece for the chapel of a hospital at a nearby Antonite monastery. The monks, who took their name from Saint Anthony, treated victims of ergotism, a painful medieval scourge, which for this reason became known as Saint Anthony's fire. Their treatment combined administering herbs and praying to Saint Anthony, who was thought to have special curative powers for the disease. Because the disease caused extremely painful skin lesions, gazing at a scene of Christ's greater agony was thought to comfort the victims.

Hieronymus Bosch's triptych, *The Temptation of Saint Anthony*, which now resides in the Museu Nacional de Arte Antiga in Lis-

Cock's spur of rye.

bon, Portugal, tells the story of the saint's spiritual and mental tor-
ments. Among other grotesqueries, it depicts an amputated foot
and a half-vegetable, half-human figure, showing another face of
ergotism, namely madness.

Fortunately, ergotism has been virtually eliminated from our
world. But we can still view the causative agent; it still infects
plants, including rye grass. The plant looks normal except for a few
of the grains in the seed head that are replaced with clearly visible
purple-black structures, sometimes called *cock's spurs* because of
their shape.

Rye was a favored cereal crop of northern Europe because it
tolerated the region's then-colder climate. The climate during the
Middle Ages was also quite rainy, which favored growth of fungi
including *Claviceps purpurea*. When ground into flour, a small
number of cock's spurs lethally contaminated large quantities of
the product with the extremely toxic alkaloid ergotamine. It is not
surprising, then, that Saint Anthony's fire was a widespread medieval horror. Gradually the association between cock's spurs and
disease was recognized and the disease was virtually eliminated.
But in 1951, a tragedy occurred in Pointe Saint Esprit, France, that
gives us a glimpse into what medieval Europe must have been like.

Five people died in this small village with 300 inhabitants; many
others suffered bizarre illness. They hallucinated that they were being chased by tigers or saw death pursuing them, and they suffered
convulsions. Some jumped from rooftops. It was not just mass human hysteria. Animals were also affected. A cat writhed, twisted,
and tried to climb a wall. A dog leaped in the air, snapped vi-
ciously, and crushed rocks in its mouth until it bled. Ducks strutted,
quacked to a crescendo, and died. The outbreak was clearly ergotism, which proved to be the result of uninspected and therefore il-
legally distributed rye flour, which was made into bread at the village bakery. The affected animals had also consumed the bread.

Ergotism has affected humans since the beginnings of agriculture.
Assyrians were acquainted with the disease as early as 600 B.C. Er-

gotism waned during Roman times because Romans did not care for rye, the favored host of *Claviceps purpurea*. But ergotism returned with a vengeance in the Middle Ages principally in Europe, where rye was the major cereal crop. Entire villages suffered hallucinations and painful, often gangrenous sores. In 994 40,000 people died of ergot poisoning in the Limoges district of France alone.

The physical and hallucinogenic consequences of ergotism are well established, as is its profound impact on human history. In addition, many plausible suspicions have been raised concerning other human calamities it might have caused or contributed to. For example, the spread of bubonic plague during the late Middle Ages and the decline of fertility that followed it might have been exacerbated by ergotism's suppression of immune function and stimulation of muscle contraction.

An even stronger case, perhaps, can be made for ergotism's impact on events leading up to the Salem, Massachusetts, witchcraft trials of 1692. About that time, wheat crops were devastated by rust (another fungal infection), so residents of Salem relied on rye for their daily bread. Women accused of witchcraft testified to having consumed sacramental bread and experienced fits, hallucinations, burning sensations, and feelings of being pricked or bitten.

One wonders also about the Great Awakening of 1741, a religious revival that occurred in New England. Thousands of people experienced fits, trances, and visions, which were then supposed to be a nervous disease. During 1889–1890, a mass occurrence of similar symptoms broke out in the Viatka province of Russia. It was accompanied by confirmed cases of ergotism.

Questions about the connection between ergotism and hallucinogenic reactions linger. Ergotamine, the active ingredient considered responsible for the symptoms of ergotism, causes muscle contraction; it has been used medically as an aid in childbirth and to staunch postpartum bleeding. It is not hallucinogenic. But it chemically resembles LSD (lysergic acid diethylamide), a well-known and extremely active hallucinogen that produces reactions quite similar to the symptoms associated with the causative agent of ergotism. Indeed, LSD was first synthesized chemically from ergotamine, the causative agent of ergotism. LSD's psychedelic properties and potency were discovered almost immediately when Albert Hofman of Sandoz Laboratories, the chemist who carried out the synthesis in 1938, touched it, thereby absorbing infinitesimal amounts of it through his skin. Extensive experiments, some done by the U.S. Army and the CIA, later established that LSD is active in microgram amounts.

A plausible explanation for the hallucinogenic or psychedelic reactions associated with ergotism is the conversion of even a minuscule amount of ergotamine to LSD or a similar compound. Some mycologists have hypothesized that other fungi that are prevalent under the damp conditions leading to outbreaks of ergotism mediate this conversion. Others speculate that the process of making bread from the ergotamine in contaminated flour is responsible for the conversion. Whatever the connections, it is well established that outbreaks of ergotism cause hallucinogenic responses in humans as well as animals.

Through monitoring, inspection, and government-imposed standards, outbreaks of ergotism have been eliminated in developed countries. But the fungus that forms the cock's spur on rye is still very much around. The spur-resembling structure ("ergot" is an old French word for "spur"), termed a *sclerotium,* that C. *purpurea* forms in the seed-bearing head of a rye plant is a hard mass of mycelium packed with lipids and other compounds, including ergotamine. The sclerotium is hardy, capable of overwintering and surviving in a dormant state for a long time. No one knows what selective advantages ergotamine offers the fungus. Certainly its lipid content and possibly the ergotamine as well contribute to its ability to survive.

In the spring, individual strands of mycelium grow out of the sclerotium and come together again in a mass, this time forming a small, mushroomlike structure in which sexual spores are produced. *C. purpurea* belongs to the ascomycete group of fungi. Some full-sized mushrooms, such as morels and *Pizizas,* are also ascomycetes, but most, including button mushrooms, delicious boletes, psychedelic psilocybes, and deadly amanitas, are basidiomycetes. The way that sexual spores (spores produced after conjugation between two mating types) are formed and arranged distinguishes the two groups of fungi. Spores of ascomycetes form inside a sacklike structure called an *ascus* ("ascus" means "sack" in Greek). Those of basidiomycetes are attached to a swollen, clublike structure called a *basidium* (which means "club" in Greek).

When windborne ascospores of *C. purpurea* land on the flowering seed head of a rye plant, they respond by mimicking a pollen grain, but with an odd twist. Those that come to rest on the stigma

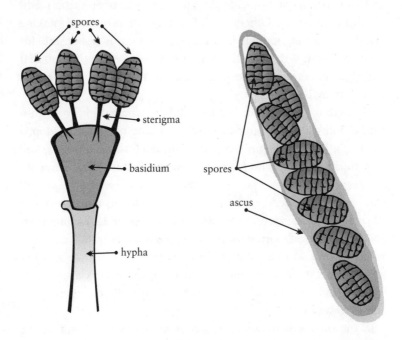

Basidium and ascus.

in the flower produce a mycelium that grows down the plant's style and enters the plant ovary as a pollen grain does. Then, instead of undergoing sexual union with the ovary as a pollen grain would, *C. purpurea* destroys it and attaches itself to the vascular bundle that would have supplied nutrients to a developing grain of rye. *C. purpurea* makes use of these nutrients as it develops into a sclerotium. At one stage of this development, a mass of soft tissue (known as a sphacelia or honeydew) forms that produces many spores capable of infecting other flowering seed heads.

As improbable as it sounds, this intricate and intriguing life cycle was completely elucidated in 1853 by the careful studies of the great French mycologist Louis René Tulasne. Knowledge of *C. purpurea*'s life cycle played a vital role in controlling the spread of *C. purpurea* and ergotism.

We turn now to more benign fungi, those that form mushrooms in quite an interesting way.

A Ring of Mushrooms on the Lawn

Often on a lawn in springtime you can see a ring of mushrooms that forms an almost perfect circle. It might be only a few feet in diameter or considerably larger—as large as sixty feet in diameter. Within a few weeks, the mushrooms get old, die, and disappear. But something remains because later, probably the next spring, a new ring of mushrooms will suddenly appear, almost overnight, in the same region, but this time the diameter of the ring will be a foot or so larger. One ring in France became almost 1,000 feet in diameter and was estimated from its rate of enlargement to be about 700 years old.

Ancients called them "fairy rings" and conjured fanciful explanations: the circle in which fairies had danced the night before, the tract where little people had ridden their horses. More plausibly today, the rings tell us something about fungi and how they grow.

Several species of mushroom form fairy rings—some reported to be edible *(Marasmious oreades)* and some poisonous *(Chlorophyllum molybdites)*. Indeed, *C. molybdites* is responsible for more poisonings in North America than any other mushroom. It causes severe gastrointestinal distress. So it is not a wise idea to rely on beguiling ring formation when selecting mushrooms for dinner.

A ring is most probably the progeny of an individual fungal spore (reproductive cell) that germinated at the center of the circle to form a mycelium, which spread underground. Because the lawn provides an even distribution of nutrients, the mycelium grew at the same rate in all directions, forming a circular leading edge. When conditions were right, the underground mycelium entered its reproductive phase by forming above-ground mushrooms. All the mushrooms in the ring are appendages of the same individual mycelium that probably developed from that single spore. The mycelium forms a ring of mushrooms instead of a solid circle of them because the live mycelium itself consists only of the leading edge. The center is dead, perhaps because it used all available nutrients and starved.

A mycelium can form a mushroom (which is really just more mycelium pressed together in a particular shape) with astonishing speed, often overnight, because it does not have to make new protoplasm (cellular contents). It just rearranges the protoplasm that it has already made. All the protoplasm in the mycelium is interconnected, forming a single branching tube because fungal mycelia have no complete cross walls. When conditions are right, the fungus simply pumps some of the protoplasm from the tubes of underground mycelium into those of the mycelium of the above-ground mushroom. Mushrooms are the fungus's transitory reproductive organs. Within them—on the gills or pores on the underside of the cap—millions of reproductive cells (spores) form, which are shed from the underside of the mushroom's cap to be dispersed by the wind. In its new home, each individual spore has the capacity to germinate and form a new mycelium, perhaps at the center of another fairy ring.

How large is an individual fungal mycelium? It can be enormous. The record holders, at least for now, are species of the mushroom-forming, tree-killing fungus *Armillaria* that are found in the northern tier of the continental United States. These fungi infect the roots of oak trees without killing them, but they are lethal to pines. In 1992, two mycologists who had become interested in the population genetics of *Armillaria* as it spread through an infected forest were astounded to discover that the samples they had taken were not from a population of *Armillaria bulbosa* (now named *A. gallica*); they were from an individual. Using sophisticated methods of DNA analysis they established that the individual fungus, arising from a single spore, covered thirty-seven acres. But unlike the fairy ring fungus, the entire area, not just the periphery, was occupied by the fungus's live mycelium. The mycelium was estimated to be 1,500 years old and to weigh more than 100 tons, about the weight of an adult blue whale. Their results, published in *Nature*, soon found their way to the popular press. The *New York Times* ran a front-page story under the banner headline, "World's Biggest, Oldest Organism—Twin Crowns for 30-Acre Fungus."

But the record did not last. Later that same year, two other mycologists reported finding an even larger individual fungus, *Armillaria ostoyae*, south of Mt. Adams in southwestern Washington. This fungus covered over 2.5 square miles. And then in 2000, an *A. ostoyae* found in Malheur National Forest in eastern Oregon broke the record again. It covered 3.4 square miles and was estimated to be 2,400 years old. Undoubtedly, there are many other enormous fungi to be found, some probably larger yet.

Curiously, it is a microbe that turns out to be the largest living thing on the planet. Our sighting of the fairy ring of mushrooms gives a hint of their remarkable lifestyle. The fairy ring apparently only decorates the lawn. It does it no harm. But not all fungi are so benign to the plants with which they associate. Some can be deadly.

Dead Oak Trees in Northern California

If you were to visit the north-coast counties of northern California today, you would note that some of the hillsides are scarred with large beige-gray areas of dying tanbark oak trees *(Lithocarpus densiflorus)* still bearing their no-longer-live leaves. Tan oak is an evergreen species, named for the particularly high tannin content in its bark and in its hazelnut-shaped acorns. Before the availability of synthetic tannins, its bark was sought after for tanning leather. In spite of their bitter, astringent taste, the oak's acorns were favored by Native Americans because they are so resistant to spoilage. Their high tannin content, which preserved them, could be removed by extensive leaching with water.

Other hillsides in northern California are even starker, peppered with empty holes occupied only with dead trees, what loggers call "snags" and Spanish-speakers call *secos.* The tan oak trees are being attacked and killed by a eukaryotic microbe, a protist, *Phytophthora ramorum,* that causes an only recently recognized disease called *sudden oak death.* Coastal live oaks as well as tan oaks are dying. Oak trees play a central role in California's ecology. They are the dominant trees in the coast range and the foothills of the Sierra Nevada. Valley oaks *(Quercus lobata)* were virtually the only trees in the Central Valley's native savanna. Certainly loss of oaks would have a devastating impact on the state's remaining natural beauty.

If such a complete microbe-caused catastrophe affected the oaks in California, it would not be the first disaster to befall a native American tree. Perhaps the most tragic infestation was chestnut blight, which eliminated the American chestnut *(Castanea dentata).*

This magnificent tree was once the most dominant species in the Appalachian Mountains from Maine to Florida and from the Piedmont to the Ohio Valley, making up about 50 percent of their hardwood forests. It was the region's prize forest jewel. Ridge tops there were so densely populated with chestnuts that they appeared to be snow-covered when chestnuts bloomed in early summer. Chestnuts were huge trees, many eight to ten feet in diameter and up to 100 feet tall. Straight grained, they were the region's most highly valued timber tree, used for everything from railroad ties to fine furniture, from crib to coffin. And the wood is said to be as rot-resistant as redwood. The nuts they produced were an added gift to wildlife and humans alike. Appalachian residents fattened their hogs and children on chestnuts. And they shipped them as a cash crop principally to New York City, where they were sold fresh-roasted by street vendors. And, as Henry Wadsworth Longfellow pointed out, they provided shade for the village smithy. Many have said you could not invent a better tree than the American chestnut.

But in 1904, disaster struck. A disease called *chestnut blight* appeared in American forests. This disease and its causative agent, the ascomycete fungus *Cryphonectria parasitica*, is native to China, where it almost benignly infects the native Chinese chestnut *(Castanea mollissima)*, causing little or no damage. At most it kills a few twigs and small branches. The European chestnut *(Castanea sativa)* is a bit more sensitive; a few European chestnut trees do die when infected. But the American chestnut proved to be tragically vulnerable. They were wiped out.

The fungus produces two kinds of spores—sexual (ascospores) and asexual spores (conidia). When either enters a wound in the tree's bark, usually one caused by an insect, it germinates, forming a mycelium that penetrates the bark, growing through it until it reaches the cambium (the tree's layer of dividing cells where the tree's girth-expanding growth occurs). Then the mycelium spreads through the nutritious cambium layer, killing it as it goes. The result is a weepy canker. When the mycelium has encircled a branch (girdled it), the branch dies. When it girdles the trunk of a tree, the tree dies.

Not all of the fungus's metabolic potential is dedicated to the mycelium spreading through the doomed tree. Part is diverted to making structures and spores so the fungus can reproduce by spreading to other trees. Certain structures, called *perithecia,* protrude through the infected tree's bark and forcibly eject ascospores, which are then dispersed by the wind, some landing on wounds in other trees. Nearby, other structures (pycnidia) form, which ooze out conidia when it rains. They are dispersed by the splashing rain or carried to other trees on the feet of birds and insects. Rain or shine, the blight can spread inexorably through a forest as it did throughout all Appalachia.

By 1950 American chestnut trees were all gone. The only remaining evidence of their prior abundant presence is the shrublike growth that comes from the roots of trees that are no longer there. These shoots, too, quickly become infected and die. Valiant efforts by the American Chestnut Foundation are under way to breed and propagate new varieties of chestnut that combine the Chinese chestnut's blight resistance and the American chestnut's majestic qualities. Some with fifteen-sixteenths American chestnut genes appear quite promising, but there is a long, hard road ahead before the restoration of the forest could even begin. Some estimate thirty to fifty years.

Another native tree, the American elm *(Ulmus americana),* has suffered an only slightly less gruesome fate from an invading fungus. Elms were probably the favorite shade tree east of the Rockies. They were common in parts of Europe and Asia as well. The elm is a near-ideal tree for lining city streets—fast growing, adaptable to many types of soil, and tolerant to the soil compaction that inevitably occurs from foot traffic in an urban setting. Its growth habit of few low branches and an upward sweeping, heavy canopy provides summer shade. Rows of the American elm formed a majestic green Gothic arch over the avenues of many American towns and cities.

The American elm's nemesis, Dutch elm disease, was caused by another ascomycete fungus, *Ophiostoma ulmi*. It, like chestnut blight's *Cryphonectria parasitica,* is an unwelcome foreign visitor. In spite of its name, which derives from the seven Dutch women scientists who studied and characterized it, Dutch elm disease most probably originated in Asia. From there it traveled to Europe in the 1910s and swept across the continent, killing 10 to 40 percent of its elms. Then in the 1940s, a curious microbe-on-microbe event occurred. *O. ulmi* itself became the victim of an epidemic, a viral epidemic. The virus did not kill the fungus, but it established itself within the fungus as a latent virus. And in so doing it markedly decreased *O. ulmi*'s ability to cause disease. The spread of Dutch elm disease in Europe virtually came to a halt.

Unfortunately, the fungus in its highly virulent form arrived in the United States before the tempering virus arrived in Europe. The fungus probably arrived on the Eastern seaboard sometime in the late 1920s, carried in infested elm timber, and was transported via these diseased logs to Ohio, where Dutch elm disease was first recognized in North America in 1930. Since then it has spread slowly but inexorably across the continent, killing over half the elms in its wake. Now it is established throughout the United States, with the exception of the desert Southwest.

Dutch elm disease is spread by bark beetles, themselves unaffected by the fungus they carry as a passive passenger. As the beetle bores into the bark, it spreads fungal spores with it into deeper plant tissues. The spores germinate, and *O. ulmi* mycelium then seeks out the tree's water-carrying vessels (xylem) and spreads through them. In response to a fungus growing in its xylem, the elm forms gum, which plugs these vital conduits. Deprived of water, first the leaves wilt and die; then the entire infected region of the tree dies. *O. ulmi,* like the chestnut blight agent *C. parasitica,* is an ascomycete fungus that produces both conidia and ascospores.

The ascospores form a sticky mass as they ooze from the neck of the perithecium where they are formed. They are the spores that attach readily to bark beetles, which when they subsequently invade another tree, start another round of infection.

Elm trees die more quickly when the fungus passes directly from one tree to its neighbor. Such transmission occurs almost inevitably between trees that are closer than fifty feet from each other, the usual situation for street trees, because elms spontaneously form "root grafts" (a merging of their tissues) when roots of one tree come in contact with those of another. Aggressive management, including removing infected branches, trenching between trees to interrupt root grafts, and removing the bark from firewood, markedly slows progress of Dutch elm disease, but it does not stop it.

Phytophthora ramorum, the microbe that causes sudden oak death, is quite a different creature from the ascomycete fungi that cause chestnut blight and Dutch elm disease. *P. ramorum* belongs to a group of microbes once called water molds and considered to be fungi, but modern molecular studies have shown that they are not closely related to other fungi. Instead they fall into that large group of eukaryotic microbes, the protists. Still, the term *water mold* is usefully descriptive. It is moldlike because it grows as a mycelium, the way fungi do. And its reproduction is dependent on the presence of water because, unlike the airborne conidia and ascospores of ascomycete fungi, its asexual spores are obliged to stay in water. They move by swimming, using their two flagella. In a way, this is good news. Sudden oak death will be restricted to the cool, damper areas of the West Coast, and it will not invade California's much hotter and drier central valley and the Sierran foothills. The not-so-good news is that *P. ramorum* has been found to have promiscuous habits. It can attack and sicken many more plant species than just tan oak and coastal live oak. Its potential victims include other oak species, California bay laurel, Douglas fir, even coastal redwood, as well as many kinds of nursery stock. In all, over a hundred plant species are susceptible to its attack. And its range is already extending to Oregon, Washington, and British Columbia. It has even made an appearance in the United Kingdom and the Netherlands.

The arrival of a microbial pathogen in a different part of the world with a more vulnerable population has been a known formula for disaster since Europeans first brought their diseases with them to America. Europeans were relatively resistant to diseases such as measles and smallpox following centuries of exposure and consequent natural selection for resistance. When they arrived in the New World, infectious disease killed an estimated 95 percent of the pre-Columbian population of Americans. In contrast, record-setting bubonic plague, which probably entered Europe on flea-infested furs arriving by newly opened trade routes from Central Asia in 1346, killed only about a quarter of the resident population. A smallpox epidemic initiated by the arrival in Mexico in 1520 of Spanish slaves from Cuba, then with a population of 20 million, is thought to have reduced the native population to about 1.6 million by 1618. The precise size of the pre-Columbian populations of Native Americans and the percentage of them killed by European diseases is controversial, but no one questions that huge numbers died.

The extreme vulnerability of the Americans caused the European diseases to spread rapidly in front of Europeans. Measles, a tolerable disease for most Europeans, proved to be a deadly killer of Native Americans. But smallpox, a feared killer of Europeans, was even more deadly to inhabitants of the New World. Smallpox killed much of the Inca society in Peru, including the emperor and his son. It had arrived before Francisco Pizarro with about 200 men conquered the disease-disabled empire. When the Spanish conquistador Hernán Cortés arrived in Mexico in 1519, the population of the region was estimated to be between 25 and 30 million. Within fifty years it had shrunk by 90 percent to about 3 million. Again, these numbers are controversial.

Native North Americans suffered equally. In 1540 when Hernando De Soto passed through the American Southeast, he came across abandoned villages depopulated two years previously by deadly epidemics. The lower Mississippi Valley, however, was still densely populated. Fifty years later, French explorers found this region similarly empty. Presumably, De Soto's just passing through was enough to start the microbial destruction of the region. Native North Americans were particularly ravaged by smallpox. Meriwether Lewis and William Clark on their 1803–1806 expedition learned from the Mandans that smallpox had almost wiped out the Arikara Indians; surviving Clatsop Indians told Clark that smallpox had destroyed their nation.

And some epidemiologists believe that such calamitous intercontinental disease transfer was a two-way street. Shortly after Columbus returned from his first voyage to America, syphilis, called the "great pox" (because of its disfiguring sores) or the French, Italian, Spanish, German, or Polish disease depending on where you were not from, appeared in Europe and spread quickly across the Euro-Asian land mass as far as China. A likely origin of syphilis is the New World disease yaws. Although causing disfiguring lesions on the skin and bones, yaws is essentially painless and nonlethal. Both yaws and syphilis are caused by species of bacteria belonging to the spirochete genus, *Treponema*. Yaws is transmitted by skin contact. Possibly it adapted to being sexually transmitted to fit the lifestyle of more fully clothed Europeans and in so doing became more virulent.

It is now increasingly clear that microbial globalization affects more creatures than just us humans. Plants and many other forms of life are similarly vulnerable to microbial disaster. Nevertheless, it is probably an inevitable part of our shrinking world's future.

Phytophthora, the genus of the fungus that causes sudden oak death, has an even more notorious member, *Phytophthora infestans*. This species caused the potato murrain, later called potato blight, that resulted in the Irish potato famine of the nineteenth century. Bubonic plague reduced the population of Europe by about

a third in the Middle Ages. The potato famine reduced the population of Ireland by two-thirds, from about 8 million to 3 million.

The murrain, as it was called, was first reported in the August
23, 1845, edition of the British *Gardeners' Chronicle and Agricultural Gazette* as a malady of potatoes in Belgium that "struck down
growing plants like frost in summer." It "consists in the gradual
decay of leaves and stems, which become a putrid mass, and the
tubers are affected by degrees in a similar way." It had been a good
spring for planting in Europe. Then in July the hot, dry weather
was followed by six weeks of cool, very wet weather. The disease
spread rapidly from Belgium to Poland, Germany, France, and then
all over England. On September 13th the *Gardeners' Chronicle* announced, "We stop the Press, with very great regret to announce
that the Potato Murrain has unequivocally declared itself in Ireland
. . . where will *Ireland* [their emphasis] be, in the event of a universal potato rot?"

Ireland at the time was a two-crop—wheat and potato—agricultural economy. The wheat was used to pay rent to absentee land-
lords and the potatoes fed the Irish farmer's own family. Potatoes
were consumed in prodigious amounts—over eight pounds per person per day. The Irish survived much of the year on potatoes and
pepper water alone. When the murrain struck, many farmers continued to pay rent as their families starved. Of course, emigration,
largely to North America, was the major reason for the country's
rapid depopulation. But microbes also exacted a terrible toll during
the voyage to North America. Ship fever, now called *epidemic ty-
phus,* a louse-born infection by a bacterium, was the major culprit.
The bacterium was named *Ricksettia prowazekii* after two microbiologists, Howard Ricketts and Stanislav von Prowazek, who died
while investigating the deadly disease.

Astoundingly, considering the infant state of the field of microbiology at the time, the riddle of the potato murrain was solved by an
amateur scientist, Reverend M. J. Berkeley, while the contagion
was still under way. He identified a fungus, *Phytophthora infestans*

(which he called *Botrytis infestans*), as being associated with the diseased plants. And he made the great intellectual leap of concluding from his studies that the fungus was not a mere scavenger feasting on a remains of diseased plant, as many fungi do. It caused the disease (although he found no way to control it). He, in fact, announced the germ theory of disease (for plants) nearly a quarter of a century before Louis Pasteur declared it (for humans).

In a way, it is curious that understanding diseases of plants preceded by such a significant interval the understanding of human disease. One would think we would have greater identification with the latter. But it is certainly true. For example, in 1861, sixteen years after the potato murrain, when Albert, Prince Consort of England (with access to world-class medical care) died of typhoid fever, the cause of the disease (now known to be the bacterium

Cryphonectria parasitica

Phytophthora infestans

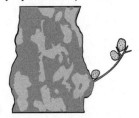

Fungal structure (stroma) emerging through the bark of a chestnut, containing three flask-shaped pycnidia, which contain ascopores that are dispersed through the necks

Fungal structures (sporangiophores) protruding from a potato leaf and bearing oval-shaped structures (zoosporangia) from which hundreds of swimming zoospores emerge

Salmonella typhi

Escherichia coli

Cryphonectria parasitica, Phytophthora infestans, Salmonella typhi, and *Escherichia coli.*

Salmonella typhi) was unknown, as were the simple hygienic and preventative measures that could have saved his life so easily. The culprit bacterium itself was not identified until 1884. And understanding viral infections came much later than that. The answer to why plant diseases were better understood long before human diseases has to do with the size and complexity of cell structures.

Bacteria such as the one that causes typhoid fever are an order of magnitude smaller than the fungi that cause most plant diseases. Viruses are one or two orders of magnitude smaller than bacteria. Moreover, the fungi are sufficiently complex in shape and structure that they can be identified and distinguished just by looking at them through a quite ordinary microscope. Bacteria cannot. For example, the causative agents of chestnut blight and potato murrain look quite different, but no one can distinguish the causative agents of typhoid fever from the laboratory workhorse *Escherichia coli* by appearance alone.

Even though fungal blights have been identified quickly, the spread of the infection often outpaces our ability to find a cure or even prevent extinction of the affected species. The dead oak trees in northern California give us a glimpse of the consequences of a new disease-causing microbe arriving in a different part of the world. Not all fast-spreading fungi are exotic foreigners, however. Sometimes, we need look no farther than behind our bathroom sink to find a fungus with a much more benign record that is nevertheless of environmental and economic concern.

Black Mold on a Wall in the Bathroom

This microbe strikes fear in the hearts of the inhabitants of the affected house. Is their health at risk? Will the value of their house or apartment plummet? How costly will cleanup and eradication be?

Many molds (fungi) are black; the one with the frightening reputation is *Stachybotrys chartarum,* which very well might be the mold growing on the bathroom wall, a highly favorable venue for it. *S. chartarum* has a fondness for the cellulose in the paper that covers wallboard, and it requires high humidity—93 percent or better. Wallboard in a steamy bathroom is just right for it. Indeed, *S. chartarum* was discovered and first described in 1837 as a black fungus growing on wallpaper in a house in Prague. The fungus is relatively easy to identify on the basis of its masses of distinctive bean-shaped, black conidia (asexual spores) that give the fungus its color and the unusual whorled cluster of cells (phialides) that bear them. *S. chartarum* does not form sexual spores. Because of its increased notoriety, *S. chartarum* is now identified by service businesses that monitor its presence using molecular methods, including PCR. No skills in mycological observation are needed.

But why is *S. chartarum,* in particular, so infamous? Several events cemented its bad reputation. In 1938, Soviet scientists solved an equine puzzle: horses in the Ukraine were suffering outbreaks of disease characterized by irritation of the mouth, throat, and nose; shock; necrotic skin lesions; hemorrhage; nervous disorders; even death. The scientists established that the tainted hay they were fed caused the disease. The hay was jet black—heavily contaminated with *S. chartarum*—and horses proved to be quite sensitive to it. The growth of *S. chartarum* from thirty petri dishes proved sufficient to kill a horse. Subsequently, such disease has been reported elsewhere in Eastern Europe, but it has not been seen in the Americas.

In the late 1930s an outbreak of disease in Russia was reported among humans on collective farms where they handled heavily *S. chartarum*–infested straw or grain, burned the straw, or slept on mattresses stuffed with it. Within a few days following exposure, they developed rashes, skin lesions, painful inflammations of the mucus membranes of the nose and eyes, tightness of the chest, cough, bloody nasal discharge, fever, headache, and fatigue. The farm workers recuperated rapidly when their exposure to contaminated forage ended. Similar episodes have been reported elsewhere in Eastern Europe.

In Cleveland, Ohio, during 1993 and 1994, a cluster of cases of pulmonary hemorrhage and hemosiderosis (iron depositing) occurred among infants. Because such outbreaks among infants are extremely rare, the story was covered widely in the press and the disease was meticulously investigated. Several events correlated with the outbreak, one being that the homes in which the infants lived had been subjected to water damage and were infested with *S. chartarum*. A study concluded that this exposure to *S. chartarum* had sickened the infants. The problem with this diagnosis, however, is that no one knew how the infants were actually exposed to *S. chartarum*. The toxicity of *S. chartarum* is attributed to compounds called *trichothecenes,* which various strains produce. These compounds have been isolated, characterized chemically, and shown to affect both animals and cultured human cells. But trichothecene toxins are not volatile. How could they have made the infants ill? The incidents in Cleveland changed completely the American public's attitude about black mold, leading many to suspect that just being around it was sufficient to make you sick. Many scientists were unconvinced by the Cleveland study's conclusions. Indeed, the Centers for Disease Control and Prevention (CDC) in Atlanta published two reports critical of the study and concluded that the link between *S. chartarum* and acute pulmonary hemorrhage/hemosiderosis in children was not proven. Of course, it is almost impossible to prove a negative, that *S. chartarum* did not cause the disease.

Despite the CDC's reports, the danger of black mold was now firmly established in the public's mind. It has resulted in multimillion-dollar litigations and caused serious problems for homeowners and building managers. *S. chartarum* has also become implicated as a significant factor in "sick building syndrome," the conviction that being in certain buildings might make you sick. An extensive review of the literature published in 2003 in *Clinical Mi-*

crobiology Review has had, for now, the final word. The authors concluded that there is no well-substantiated evidence linking the mere presence of *Stachybotrys* species to illness. Certainly, black mold on the bathroom wall is unsightly and it ought to be removed, but it is no basis for excessive fear or panic.

10

Viruses

As we have already discussed, viruses have only a tenuous claim to being microorganisms. But they do have enormous impact on cellular creatures, all of which are vulnerable to viruses. The vast majority of cellular microbes are environmentally beneficial. Some play vital, irreplaceable roles keeping Earth habitable. Only a minuscule minority of cellular microbes harms us or other creatures. But the same cannot be said for viruses. Their possible positive impact, if it exists, remains unknown. Nobel laureate David Baltimore put it unequivocally, "If they weren't here, we wouldn't miss them." The damages viruses cause their hosts are manifold. Another Nobel laureate, Peter Medawar, famously emphasized their fundamentally pernicious nature, describing a virus as "a piece of bad news wrapped in a protein."

Virus-caused diseases have long devastated humans, plants, and animals. But progress has been made. Smallpox, possibly the greatest killer of humans, was eliminated worldwide in 1979 by vaccination, and many other deadly viral diseases, including measles, poliomyelitis, rabies, and yellow fever have been prevented or controlled by vaccination or other public health measures, at least in the developed world.

It's a continuing struggle, however. New viral diseases emerge relentlessly. AIDS, caused by human immunodeficiency virus-1 (HIV-1), appeared in the United States in about 1970. Severe acute respiratory syndrome (SARS), caused by SARS coronavirus, was unknown before 2003. Dengue virus, West Nile virus, monkeypox virus, Ebola virus, and H5N1 avian influenza lurk as potential perpetrators of catastrophic pandemics. Yearly, new forms of seasonal influenza, each necessitating a new vaccine, wreak suffering and death.

Viruses also target plants and livestock. In the United Kingdom an outbreak in 2001 of foot-and-mouth disease brought major economic distress to the beef industry. Plum poxvirus, which kills stone fruit trees, has recently appeared in the United States and Canada. Colony collapse disorder, responsible for the disappearance of many honeybees, probably has a viral connection. Protecting ourselves, plants, and animals from viral attack is a never-ending challenge. In this chapter we examine some of their notable attacks on animals.

A Skunk Staggering across the Trail at Midday

Unless something is terribly wrong, the graceful, dignified, nocturnal skunk does not venture out in the daytime, and it certainly does not stagger. This poor skunk is most probably suffering from rabies, a viral disease. It is a good idea to step back and give him the right-of-way. His disease has invaded his brain, fundamentally changing his behavior. Instead of cautiously relying on his tremendous powers of chemical deterrence, he is likely to be aggressive and ready to bite. And his saliva contains infectious virions.

Unlike most viruses, which attack only one or a few species, rabies is an equal-opportunity pathogen, able to infect any mammal (although some are particularly susceptible), and its effects are horrific. Hydrophobia (fear of water), the disease's once-common des-

ignation, is one of its classic symptoms. Because the muscles controlling swallowing contract abnormally, the patient cannot drink.
When a mouthful of water is taken, the victim either chokes vio-
lently or spits it out in an uncontrollable spasm—certainly an experience that would make one deathly afraid of water. Hydrophobia
is only one of rabies' neurological horrors. Others include loss of
speech and vision, as well as extreme apprehension and anxiety.

Transmitted principally by animal bites, rabies virus multiplies
in the tissues near the wound. Eventually, it infects a nerve cell
and travels through it to the brain, where it exerts its perception-
changing and eventually deadly consequences. Once the disease has
developed, no treatment is possible, and it almost certainly kills.
The period between bite and the development of symptoms as well
as the probability of developing disease depends to a considerable
extent on the closeness of the bite to the brain, and it varies enor-
mously—from a little over a week to several years. During that period, rabies can be prevented by immunization. Louis Pasteur developed such a vaccine in 1885. Now more effective and less toxic
vaccines are available. They are taken preventatively by people such
as veterinarians, who are much more likely to be bitten. Most
places in the United States require dogs and cats to be vaccinated, a
precaution that dramatically lowers the incidence of rabies in humans. Now fewer than a dozen cases of rabies are reported annually in the United States. Some places, notably the British Isles, are
rabies free.

Some aspects of rabies virus—such as its ability to infect a wide
variety of animals—are unusual. Not many viruses can be thwarted
by vaccine therapy after the virus has entered the host. But other
characteristics are quite typical of viral infections, most notably its
not being treatable by antibiotics or other chemotherapeutic agents
once the infection is under way. That is because rabies and most
other viruses have no targets that the host cell lacks; they cannot
because viruses use the host cell's own metabolic machinery to replicate. So treating viral infections with antibiotics is an exercise in

futility. In fact it is a bit worse than that because such unnecessary administration selects for resistant strains of bacteria that can share their newly acquired ability with other bacteria and initiate a resistance cascade, leading to public health disasters. Rare antibiotic-resistant cells that are always present survive, flourish, and become dominant in the antibiotic-laden environment. They share their resistance genes with other bacteria. The consequences include the prevalence of newly arrived killers such as methicillin-resistant strains of *Staphylococcus aureus* (MRSA) and even more ominously a vancomycin-resistant strain of *S. aureus* or multiple drug-resistant strains of *Mycobacterium tuberculosis*.

HIV (human immunodeficiency virus), which causes AIDS (acquired immunodeficiency syndrome), is a dramatic exception. It can be successfully treated with drugs, although not cured, even after it has initiated an infection. The reasons why give us insight into this virus's unusual lifestyle.

HIV belongs to a group of viruses called *retroviruses,* so named because their genetic information flows in a reverse direction. The usual way genetic material flows in cells, which is so prevalent as to be called biology's central dogma, is from DNA in the cell's genetic repository to RNA as its genetic signals are being expressed and then on to protein. But retroviruses' genetic information, including HIV's, initially flows in the opposite (retro) direction.

In the virion, HIV maintains its storehouse of genetic knowledge as RNA. Lots of other viruses do that, but unlike HIV, when they infect a cell they let the stored information flow directly to protein synthesis, as cells do in the last part of the pathway of normal genetic flow. These other RNA viruses short-circuit the genetic flow, DNA to RNA, of cells. They do not reverse it. In contrast, the first thing HIV does is reverse the flow—from RNA back to DNA. Then the DNA inserts itself in the host cell's genome. Only later does the usual flow begin: viral DNA to viral RNA to the viral proteins that do their nastiness.

The host cell has no machinery to mediate HIV's RNA-to-DNA step, so the virus itself must provide it. It does so by incorporating the mediating enzyme, *reverse transcriptase,* right in its own virion. (This enzyme is encountered elsewhere in nature, but such occurrences are unusual.) Reverse transcriptase, which is essential for HIV replication, offers an ideal target for anti-HIV drugs because our cells do not have or need it. A whole set of such drugs with now-familiar names such as AZT has been developed that inhibit this enzyme. They are largely responsible for changing fundamentally the prospects of those infected with HIV from imminent death to merely dealing with a difficult, lifelong disease that is expensive to manage. For reasons that are still far from clear, these anti–reverse transcriptase drugs arrest the development of HIV, but they do not eliminate it. They extend healthy life. They do not cure the disease.

Owing to another aspect of its unusual lifestyle, HIV offers an-

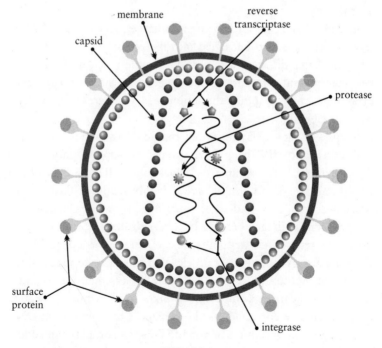

HIV virion.

other target for anti-HIV drugs. When an HIV virion is shed from an infected cell, it is still incomplete: several of the viral proteins within it are linked together in a long strand that must be cut into individual proteins before they become active. Such final processing is brought about by yet another enzyme, which is unique to HIV and is packaged within the virion where it is needed to divide the precursor protein into its active components.

This second viral enzyme is HIV protease. *Protease,* as I mentioned briefly in Chapter 3, is a general term for enzymes that cut the chemical bonds holding amino acids together in proteins. So there are lots of proteases in nature, including in our own cells. But HIV protease has special properties that make it a unique viral target that anti-HIV drugs can attack and inactivate without damaging the host cell. Some of the most effective anti-HIV and therefore anti-AIDS drugs, protease inhibitors such as Indinavir or Norvir, act this way.

And HIV offers yet a third potential target, an enzyme called *integrase,* which is also packaged into the mature HIV virion. Integrase mediates the step in the HIV life cycle when its DNA form becomes integrated into the host cell's genome. Successful anti-HIV drugs that act by inhibiting the action of HIV integrase have not yet been developed.

In spite of their delaying rather than curative actions and their high costs, anti-HIV drugs have given new hope to patients and doctors in the battle against the devastating AIDS pandemic. But even these marvelous drugs have a highly tentative future. Like other microbes, HIV undergoes mutations, some of which lead to drug resistance. In fact, because HIV goes through a phase in which its genetic records are written exclusively in the form of RNA, the virus undergoes mutations at an elevated rate. The accuracy of the process of replicating RNA is considerably less than the accuracy of DNA replication. Moreover, mistakes made while replicating RNA cannot be corrected, as those made while replicating DNA can. So, even in the course of an individual person's HIV infection, multiple drug-resistant forms of HIV are sure to develop.

Multiple drug therapy is the preferred counterattack. The rationale is simple enough. Although resistance to a single drug is highly probable, simultaneous resistance to several becomes extremely unlikely. For example, if drug resistance occurs at a rate of one in a million replications (an alarmingly high frequency), only one in a million of those one in a million copies will be simultaneously resistant to two drugs, i.e., one in a trillion, somewhat improbable. And simultaneous resistance to three drugs would be one in a million trillion. That is getting highly improbable. But as we have all learned, lots of things can go wrong. The period during which the AIDS pandemic can be held in check by existing anti-HIV drug therapy is undoubtedly limited. Prevention is always best.

These two especially pernicious viruses, rabies and HIV, introduce us to the complexity of viral replication and their potentially devastating effect on animals and on us. But what if a species of animal, unintentionally imported, was the source of an ecological disaster? Could a virus be used to mitigate its impact? This was the question in Australia.

Rabbits in Australia

About a hundred million rabbits live in Australia today. Each rabbit's existence is residual evidence of one of the most thoroughly studied interactions between a population of mammals and a deadly pathogenic microbe to which it has been newly exposed. The consequences of this interaction have shed considerable light on how humans and pathogenic microbes coevolve following their first encounter, either when the pathogen is introduced into a new population or when a new pathogenic microbe emerges.

Before Europeans began to settle there in the late eighteenth century, Australia, then New Holland, the land of marsupials and monotremes, was free of placental mammals, except for its aboriginal inhabitants and their dingo dogs. Europeans brought with them their own animals—mammals and birds—for farming, amusement, and sentimentality. Among these were rabbits. In 1859 an Englishman, in order to provide a "spot of hunting" and "a touch of home" imported twenty-four European rabbits, five hares, seventy-two partridges, and some sparrows. The rabbits prospered almost microbelike in Australia's grasslands, reproducing at an alarming rate. After only six years, there were 22 million rabbits in Australia. Their range expanded virtually as fast—about seventy miles per year, reaching every part of the continent less than fifty years later in 1907. By the 1930s the rabbit population was estimated to be a staggering 750 million. Like the growth of a microbe in a favorable environment supplied with adequate nutrients, the growth of Australia's rabbit population was exponential: the rate of increase as well as the actual numbers grew constantly.

Rabbits soon became a plague, decimating the grass forage of native wallabies and domesticated sheep, causing widespread erosion of the soil, fouling water holes, consuming crops, and killing young trees by eating their bark. Conventional control measures such as shooting, trapping, or poisoning were powerless to stop their relentless expansion. Even the unconventional approach of building thousands of miles of rabbit-proof fences across the entire continent failed to contain them.

The somewhat desperate Australian government turned for help to a microbe, *Myxoma* virus, which is native to South America. *Myxoma* causes only a mild disease when it infects the local rabbits (species of *Silvilagus*), a disease characterized by inconsequential skin lesions. But it causes a deadly disease, called *myxomatosis,* when it attacks European rabbits *(Oryctolagus cuniculus),* forming readily visible, grotesque lumps. The disease progresses to causing puffiness around the head, conjunctivitis, possible blindness, listlessness, loss of appetite, and fever. Frequently a secondary bacterial infection sets in. Death follows, usually within thirteen days.

In 1950, *Myxoma* virus–infected rabbits were released in Australia with the aim of starting an epidemic. It certainly worked. The virus, carried by mosquitoes, spread rapidly and lethally. Within

two years, the continent's three quarters of a billion rabbit population had been reduced to 100 million. Almost all (over 99 percent)
of infected rabbits died. One would think that the rabbits of Australia were headed toward extinction. But that is not what happened. Microbial epidemics rarely, if ever, cause extinction, even
when a completely sensitive population is exposed to a highly lethal pathogen, because there is intense selective pressure on both
host and pathogen to evolve toward greater mutual tolerance. In
the case of myxomatosis, survival time of the rabbits increased and
lethality progressively dropped to its present approximately 25 percent. (Recall that *Myxoma* virus's killing rate of rabbits in South
America, where host and pathogen have coexisted for at least many
centuries, is vanishingly low.)

With the decline of lethality of myxomatosis, the Australian government sought new ways to control their again-rising rabbit population. A second rabbit-killing virus (rabbit calicivirus) was introduced in 1996. The outcome of this massive experiment is still in
doubt.

The selective pressure on the host for greater resistance to viral
attack is obvious. Those hosts with greater intrinsic ability to survive pass on these advantages to their progeny. But there is also selective pressure for moderation on the pathogen. The pathogen is
dependent on the availability of hosts for its own survival. There is
no selective pressure for a pathogen to kill its host. A longer-lasting
infection (longer survival of the host) during which the pathogen is
available to be transmitted to another host also benefits the pathogen's dissemination and survival as a species. A well-adapted pathogen causes a long-lasting, benign infection. Newly emerging pathogens or established pathogens introduced to a new population of
hosts tend to be highly lethal. Then, with the force of powerful selective pressures, they inexorably evolve toward mutual tolerance.
There are many examples in the history of our own species' exposure to new pathogens.

The human bacterial disease syphilis *(Treponema pallidum)* is certainly not a trivial infection. It causes initial genital sores, which go away. The disease then progresses slowly, leading after many years in most cases to the death of an untreated victim. But in 1495 when syphilis first appeared in Europe, most probably from the New World, it was a horror deservedly called "the great pox." Sores covered the body from head to knees. Flesh fell off victims' faces. Death followed infection by only a few months. Only fifty years later, however, the disease had evolved to the less horrible and longer-lasting one we know today.

We have seen similarly accommodating changes occur in tuberculosis (TB) during the past century. Tuberculosis, the white plague, caused by the bacterium *Mycobacterium tuberculosis,* was almost the defining disease of the nineteenth and early twentieth centuries. Its victims were many—literary (Emily and Anne Brontë, Elizabeth Barrett Browning, Lord Byron, Robert Louis Stevenson, Henry David Thoreau, George Orwell, Ralph Waldo Emerson), musical (Frédéric Chopin, Niccolo Paganini), political (Simón Bolívar, Andrew Jackson, Eleanor Roosevelt), even fictional (the character of Mimi in Puccini's *La Bohème*)—among many others less well known. Since the nineteenth century the incidence of the disease has declined at a steady rate of 1 to 2 percent per year until the present. Today it remains a dreaded disease but kills relatively few people. In the United States, for example, 711 people died of TB during 2004 as compared with 654,092 from heart disease and 550,270 from stroke.

Of course, many factors, including better living conditions, improved nutrition, and the availability of anti-TB drugs, opposed the disease, but the most significant factor is the bacterium's own decline in virulence. The introduction of anti-TB drugs, for example, can hardly be detected as a blip on the curve tracing its rate of decline.

The host, as I noted, also evolves toward greater resistance as the pathogen evolves toward doing less damage. In many cases it is difficult to determine the relative contribution of these two factors.

The arrival of European diseases in the New World, however, offers unambiguous evidence of the importance of the host's genetically determined resistance. They were the more-resistant survivors and descendants of survivors of bouts with these microbes. Native Americans had never been so challenged. The microbes were much more lethal to Native Americans than even the Europeans themselves.

New, highly destructive diseases are introduced to vulnerable populations by other means as well. One of these is by "jumping" a species barrier. The most well known of these species-jumping pathogens is HIV, which causes AIDS. HIV jumped the species barrier from monkeys to humans. Influenza virus is another example of such a pathogen; its native home is most probably ducks. Perhaps HIV will also moderate as other diseases have, but influenza is a different story. New strains of the virus that evolved in various birds or swine keep jumping to our species.

Whether a virus is introduced intentionally to cause a disease, as with myxomatosis, or whether the disease leaps species or geographic boundaries, as with HIV and influenza, the initial consequences can be deadly. In our next microbe sighting, we will see how attempting to prevent such effects can cause its own consequences.

Rabbits in Spain

Rabbits do not thrive everywhere. The Spanish rabbit population is relatively small and Spaniards would like to protect rather than eliminate them. Because Spanish rabbits are European rabbits, they are quite sensitive to myxomatosis, and the Spanish set about ways to protect their rabbit population from this deadly disease. They decided on vaccination. Vaccination, however, has some limitations.

The European attitude toward wild rabbits and whether or not to control them varies by nation. Britain was rabbit-free until they were imported from North Africa in the twelfth century. The populations of rabbits in Europe remained moderate and stable until the early twentieth century when they skyrocketed, presumably because of changed agricultural practices. Then rabbits became a serious pest, and farmers employed a variety of methods to control them. Remember Mr. McGregor? None of the measures had much impact.

Rabbits were also pests in France. During the summer of 1952 a retired physician introduced myxomatosis to control the rabbits, which were plaguing his estate near Paris. Of course, the disease spread rapidly beyond the boundaries of his estate and beyond France, reaching Britain by the fall of the following year. (Although mosquitoes are the vector that spreads myxomatosis in Australia, rabbit fleas do so in Europe.) As was the case in Australia, the slaughter was almost but not quite complete. An estimated 99 percent of the rabbits in Britain died. However, again rabbits and virus coevolved toward moderation. The rabbit population of Britain is now about half what it was before myxomatosis arrived.

Myxomatosis also spread to Spain, endangering their sparse population of rabbits and the animals that prey on them as well. Spanish microbiologists have developed a carefully engineered vaccine to control the virus that other nations have employed to control rabbits. The vaccine is a live *Myxoma* virus still capable of multiplying in rabbits (and thereby conferring immunity) but attenuated, so that it no longer elicits significant disease. In addition, the vaccine carries a gene for a protein component of rabbit hemorrhagic disease virus (RHDV) so that it confers immunity to this rabbit-killer as well. Moreover, the virus has been genetically debilitated so that it can survive only a single rabbit-to-rabbit passage. The vaccine sounds ideal. Because it is live, it is infectious and spreads through a rabbit population just as wild-type *Myxoma* virus does, but it protects rather than kills. Being debilitated, its spread is restricted to the immediate area where it was released. The vaccine is being tested for efficacy and safety on a small island.

Of course use of such a live, self-spreading vaccine is controversial. Were it to mutate toward greater ability to spread (and natural selection would favor such a mutation), the utility of *Myxoma* virus and RHDV would be greatly compromised. Concern about the unintended consequences of releasing viruses, naturally occurring or engineered, is certainly well established.

Initially all vaccines, whether against cellular or viral microbial pathogens, were live. The first vaccine, the one developed by Edward Jenner in 1796 to control smallpox, was a live preparation of a related virus that caused the mild disease cowpox. (This inspired Jenner to coin the term *vaccine* based the Latin word *vaca* for "cow.") Milkmaids who commonly acquired cowpox and never seemed to become victims of smallpox developed crusty sores on their hands. The scabs were ground and used as the vaccine, which proved highly effective. Cowpox is still used to protect against smallpox. Its protective power was the means by which smallpox was eradicated worldwide in 1979. Stocks of the virus remain in the United States and Russia, and perhaps unofficially in other places, so there is a certain fear that it will return. In view of this threat, some believe that prophylactic vaccination for smallpox should be reinstituted.

In the nineteenth century, a different sort of live vaccine, attenuated vaccine, was developed by Louis Pasteur. Such vaccines provided an established route of controlling infectious disease. The rationale for their development was this: cultivate the actual disease-causing microbe under suboptimal conditions with the hope that the microbe, by adapting to these unfavorable conditions, will be weakened and lose its ability to cause disease (its virulence), but retain its ability to stimulate immunity to that disease. Pasteur used this approach successfully to develop protective vaccines against the bacterial disease anthrax and the viral disease rabies.

Developing attenuated vaccines was problematic because it was unclear what would weaken the microbe; finding ways to attenuate was very much trial and error. Pasteur attenuated the anthrax bacillus *(Bacillus anthracis)* through prolonged cultivation at a temperature above its optimum. He had no idea why this treatment caused anthrax to lose virulence, but it did. Now, the mechanism is clear: when bacteria, including anthrax, are grown at elevated temperatures, they tend to lose their plasmids. The genes encoding anthrax toxin, which causes the disease, are carried on a plasmid. Pasteur attenuated rabies virus by multiple passages through rabbit brains. The mechanism for such attenuation remains a mystery.

The next advance in vaccine making was the use of killed-microbe vaccines, which greatly simplified the process. The disease-causing microbe is simply killed and used as a vaccine. It cannot cause disease and retains some immune-conferring abilities. But there are disadvantages. In general, killed vaccines are not as effective as live vaccines because they cannot multiply. With many such vaccines, booster inoculations have to be administered to achieve and maintain adequate protection.

Both attenuated- and killed-virus vaccines have been used in the battle against polio—an undertaking that is rapidly approaching worldwide victory. The first polio vaccine, often referred to as *injectable polio vaccine* (IPV), the one developed by Jonas Salk, is a killed-virus vaccine. The later one, called *oral polio vaccine* (OPV), developed by Albert Sabin, is an attenuated-virus vaccine developed by passing the virus through suckling hamsters many times.

The development and widespread use of vaccines has exerted a monumental impact on public health, probably greater than that of the advent of antibiotics. It has virtually eliminated the fear of such formerly horrifying diseases as diphtheria, polio, yellow fever, and rabies. But like all great medical advances, there are downsides. Diseaselike side effects are one problem. Such side effects can be minimized by using purified or acellular vaccines that contain only

11

Felonious Bacteria

I have thus far emphasized the many ways that bacteria help us or our planet. But of course bacteria have no notion of working for the common good. In this chapter we investigate some of their harmful impacts, which occasionally do some good as well.

A Friend Who Looks Just a Bit Younger

There could be a lot of reasons your friend looks a bit younger and fresher today, but Botox, a highly dilute preparation of microbe-produced toxin, has to be one possibility, indeed an increasingly probable one. This terribly toxic killer of millions does, indeed, remove wrinkles, at least for a few weeks or so.

Botox and Dysport are trade names for botulinum toxin, one of the most lethal substances known—a dubious honor, to be sure. It competes for the title with two other deadly microbial products: diphtheria toxin and tetanus toxin. Botulinum toxin is about 10,000 times more toxic than strychnine. Around a quarter of a pound of it, evenly divided, would be sufficient to kill all humans on the planet.

the microbes' immune-stimulating proteins, not those that cause unpleasant side effects.

Another limitation for controlling disease by vaccination comes from microbial trickery. It has not proven possible, for now at least, to make effective vaccines against certain notorious microbial pathogens, for example *Neisseria gonorrhea* (gonococcus), the causative agent of the sexually transmitted disease gonorrhea, because this pathogen has the capacity to change its surface proteins frequently, rapidly, and randomly. To protect us from disease, our immune systems recognize these surface proteins and make protective antibodies. But if the pathogen's surface proteins are always changing, our immune system can't keep up. Gonococcus changes its surface proteins by switching their encoding genes on and off. Just as the host's immune system becomes able to make protective antibodies against gonococcus's surface proteins, the invader switches those encoding genes off and turns a new one on. The immune system has to start all over again. This cycle can repeat dozens of times.

Other microbes engage in similar trickery but by different means. The surface antigens of some microbes such as influenza change as their encoding genes change, either incrementally as a result of mutations (a process called *genetic drift*) or precipitously, as influenza exchanges its complement of chromosomes with another strain of influenza during simultaneous infection of the same cell (a process called *genetic shift*). Other microbial pathogens, notably *Plasmodium* species (which cause malaria) and HIV, have resisted being controlled by vaccination owing to the complexity of their lifestyles.

Vaccination, though not a panacea, has proven to be a powerful weapon in the fight against microbes that would harm us and other animals, as the rabbits in Spain might attest. In spite of vaccines' many invaluable contributions to human welfare, doubts and controversies still surround their use. Control of diphtheria is notable. This deadly bacterial disease, caused by *Corynebacterium diphtheria,* has been a known human killer since Hippocrates described it clinically in the fourth century B.C. Its horrors continued world-wide well into the twentieth century. The throat swells and becomes covered with a grayish pseudomembrane. Together swelling and a pseudomembrane can obstruct the airway, causing sudden death. Diphtheria's old Spanish name was *garrotillo,* strangling disease. It was an ever-present threat. In the United States in the 1920s, when the first reliable data were collected, approximately 150,000 cases and 13,000 deaths were reported annually, equivalent to about 37,000 deaths for today's population, which is approximately equal to the number of traffic fatalities in the United States in 2008 (39,800). Then in 1929, Gaston Ramon, working in the Pasteur Institute of Paris, developed an effective vaccine made by inactivating diphtheria toxin, the protein that on its own can cause the symptoms of diphtheria. As the vaccine's acceptance and use spread, diphtheria declined precipitously. In the United States, where diphtheria vaccine is given to infants during their first six months of life as a component of DTP (diphtheria-tetanus-pertussis) vaccine, only five cases were reported between 2000 and 2007. Pediatricians who kept a long fingernail ready in an emergency to clear the pseudomembrane from the throats of gasping diphtheria patients were succeeded by those who never encountered the disease.

Diphtheria is only held in check by vaccination. It has not been eliminated. Vaccinations declined sharply with the chaos accompanying the fall of the Soviet Union. In 1998, according to Red Cross estimates, there were 200,000 cases of diphtheria in the territory, with 5,000 deaths, causing the Guinness Book of World Records to list diphtheria as the "most resurgent disease."

Still, some vehemently oppose vaccination, believing their children are "over-vaccinated" or that vaccination is responsible for the rising risk of autism. A mercury-containing preservative (thimerosal) added to many vaccines since the 1930s was a suspected culprit. Although there is no evidence that thimerosal is harmful, other than possibly causing redness at the site of inoculation, in 1999

government agencies, medical societies, and manufacturers agreed on the precautionary action of reducing the amount of thimerosal in vaccines or eliminating it completely. Since 2001, with the exception of some influenza vaccines for adults, thimerosal is not added to vaccines. No vaccine to be administered to children contains thimerosal. In spite of these precautions, there has been no diminution in the incidence of autism.

Refusing vaccination presents a moral conflict. In most cases in the developed world, refusal does not materially increase the child's chance of disease, but it is a hazard to society. The phenomenon of herd immunity protects nearly all of us. In a society in which 75 to 80 percent of the population is vaccinated for diphtheria (the percentages vary by disease), the unvaccinated are protected because the vaccinated break the chain of infection. If, however, vaccination incidence were to fall below 75 percent, the consequences could be catastrophic for the unvaccinated.

Botulinum toxin is made by *Clostridium botulinum,* inelegantly named for the food in which the poison sometimes lurks (*botulus* is Latin for "sausage"). *C. botulinum* is among the minority of microbial bad actors that do not contribute to keeping our environment in harmonious balance but do sicken or kill us. But we are not singled out; it also sickens and kills huge numbers of wildlife, notably waterfowl.

C. botulinum differs from most microbial killers in that it rarely infects us by invading our bodies, growing there and causing disease as most pathogenic bacteria do. Usually it merely produces its lethal toxin wherever it may be living. We become sick or die when we consume its toxin-laden surroundings. Two of *C. botulinum*'s characteristics render it particularly treacherous: it is an anaerobe, and it produces endospores. Being an anaerobe means that it can grow (and produce its toxin) in oxygen-free places such as canned goods or deep within that namesake sausage, for example. Producing endospores means that it can survive extremely harsh conditions.

Endospore-forming bacteria produce these survival forms when they run out of nutrients. Endospores do not need nutrients to survive because they are metabolically inert, showing no other signs of life than the ability to germinate and become ordinary bacterial cells again when conditions improve. And they are extremely tough. As we noticed in a previous sighting, they can survive hundreds of years in their inert state as well as hours in boiling water. Exposure to temperatures above 250°F for fifteen minutes is needed to kill them, a condition that can only be attained in a pressured vessel. Indeed, industrial standards of canning food are designed to kill endospores of *C. botulinum.* If they are killed, all other organisms will die as well. Nothing else can withstand such treatments.

Not so long ago, home canning was a common way to save some of summer's garden bounty. Unfortunately, it was occasionally a notorious source of botulism. Tomatoes are safe because surviving endospores cannot germinate and grow in acid media. But neutral foods such as green beans can harbor the deadly toxin. When opened, the contents look perfectly okay, but even a small taste of it can kill within a day or so. Double vision progresses to blurred vision, drooping eyelids, slurred speech, dry mouth, muscle weakness, paralysis of muscles, including those needed for breathing, and possible death. Now only about 110 cases are reported annually in the United States, and they can be treated with mechanical assists for breathing along with intensive medical and nursing care, which allows the patient to survive until the effects of the toxin wear off.

This disease is called *botulism,* a foodborne illness. It is an intoxication, an impact of a toxin produced by a microbe, not caused by the living activities of the microbe itself. But foodborne botulism is not the only havoc wreaked by *C. botulinum.* Two other kinds—wound botulism and infant botulism—which are true infections, also occur. Wound botulism is a consequence of *C. botulinum*'s ability to grow in an anaerobic, nutrient-rich wound. Botulinum toxin produced there enters the blood, causing the same consequences as ingesting the toxin in food. Infant botulism, which accounts for about three-quarters of cases of botulism today, is a consequence of *C. botulinum*'s ability to grow in an infant's intestines. Infants from five to twenty weeks of age, whose intestinal tracts have not yet established a competing population of microbes, are susceptible. When *C. botulinum* grows there, it produces toxin, with the same results as taking the toxin in food. So likely sources of *C. botulinum* are a hazard to infants. Honey is a rich source. That is why pediatricians recommend against feeding honey to infants.

In spite of its deadly consequences, botulinum toxin is made up of just two rather ordinary proteins joined together by easily broken chemical bonds—sulfur-to-sulfur bonds between cysteine amino acids in the two proteins. The toxin's lethal strike is fiendishly precise. Its heavier protein component is endowed with the capacity to seek out and find the exact spot where nerves contact muscles in order to energize them. When botulinum toxin arrives at this critical junction, the lighter protein, which is a protein-destroying enzyme (a protease), attacks the neuromuscular junction, destroying it so that the neurotransmitter, acetylcholine, cannot bind to it. Without the intervention of acetylcholine, muscles become flaccid, and the nasty consequences of botulism poisoning ensue. One of *C. botulinum*'s very close relatives, *Clostridium tetani,* the microbial bad actor that causes tetanus (also called "lockjaw"), produces a similar toxin that has quite the opposite impact on muscles. It causes them to become rigid.

You might suspect that botulinum toxin would be a candidate for use in biological warfare because it is so potent. Fortunately, it is readily inactivated by oxygen and sunlight, so it is not a good candidate. Instead, it has found its way into beauty salons. But its use for terrorism of various sorts has been considered. The Aum Shinrikyo cult, which planned and carried out a deadly attack on Tokyo's subway system, apparently considered releasing botulinum toxin because they built facilities to manufacture it. Instead they released sarin, a deadly organophosphate and nerve gas developed in Germany as a pesticide. Some even claim that in 1961 the CIA dipped some of Fidel Castro's favorite brands of cigars in botulinum toxin to prepare for an assassination attempt.

Botulinum toxin is an ingredient in the spa treatment Botox, which worked its way into beauty salons and spas via the medical wards. Being a powerful muscle relaxant in very small doses, it has valid medical applications such as for treatment of spasticity and muscle pain. Other possible uses include treatment of migraine headaches, prostate symptoms, asthma, and obesity. In all cases, the muscle-relaxing activity of very small doses of botulinum toxin was presumed to be helpful when prescribed and administered by a medical professional. In April 2002, Botox was approved by the FDA for cosmetic applications, and that has become botulinum toxin's major use. Injecting it near wrinkles relaxes the surrounding muscles, causing the wrinkles to disappear. Such treatment is generally agreed to improve the appearance of frown lines between the eyes as well as necklines, to make one appear to be younger as your friend did since last you saw her. She did have more frown lines before, did she not?

A Crying Child Pulling on His Ear

The child's pain is probably intense, and the parents' distress probably equally so. No doubt the child has an earache, and it hurts. But the cause is not quite so obvious. Almost certainly, the child's middle ear is filled with fluid and the eustachian tube, which should drain it, is plugged. Pressure has built up in the middle ear, causing the pain. The condition is called *otitis media,* inflammation of the middle ear.

It could be caused either by an allergy or an infection. Typically, a sudden-onset, painful earache (acute otitis media) is an infection that follows a cold or a sore throat. The middle ear cannot be viewed directly because it is hidden behind the eardrum (tympanic membrane), but evidence of the pressure that builds within it can be assessed by looking at the eardrum from the outside through an otoscope (that lighted tube with a magnifying lens that doctors stick in your ear canal). Pressure in the middle ear causes the eardrum to bulge outward.

Otitis media is common in infants and children because their immune systems are not yet fully developed and their eustachian tubes do not yet drain down as those in adults do. Seventy-five percent of us have suffered at least one ear infection by the age of three.

But what action should be taken? The primary symptom is pain, and that certainly should be treated with a painkiller on your doctor's recommendation, probably one available over the counter. But what to do about the underlying infection is not so clear. If it is a bacterial infection, the culprit could be pneumococcus *(Streptococcus pneumonia),* the bacterium with the thick capsule that guided

Frederick Griffith to the discovery of the transforming principle, and then Oswald Avery, Colin Macleod, and Maclyn McCarty to the monumental discovery that genes are made of DNA. Or it could be either of two gram-negative pathogens, *Haemophilus influenza* or a species of *Moraxella*. Should antibiotics be given? Now, most pediatricians agree to treat with antibiotics only under certain circumstances: if the child is less than two years old or running a high fever—over 102.2°F. Then the dangers of life-threatening infection are quite real. But otherwise the infection is self-limiting; treated and untreated patients recover equally rapidly.

The hazards of antibiotic treatment are both individual and collective. Although an antibiotic such as penicillin is no more toxic than table salt, the treated individual can suffer severe intestinal upset because his or her complement of intestinal bacteria is dramatically altered by the antibiotic. There is also a certain hazard that the child may become allergic to penicillin, and therefore will not be able to use it at a later, more critical time. Our collective hazard of unnecessary use of antibiotics is the development of resistant strains of deadly pathogens. These strains are unaffected by one or several antibiotics, rendering these remarkable life-saving drugs—aptly called "wonder drugs" when they were first discovered—essentially useless.

Penicillin, the first antibiotic to be discovered, fulfilled Paul Ehrlich's dream of an ideal chemotherapeutic agent—a "magic bullet" that hits and kills the invading microbe without damaging us, the microbe's vulnerable host. Ehrlich, a Nobel laureate who died in 1915 before the discovery of penicillin in 1929, was considered the "father of chemotherapy" because of his discovery of the first chemotherapeutic agent, salvarsan, which can cure syphilis. But salvarsan fell short of "magic bullet" status: it's quite toxic to humans. Penicillin was the answer. It hits and kills a vital target in the bacterial cell that our cells do not even contain. The target is peptidoglycan, or more accurately, murein, a form of peptidoglycan that occurs in cell walls of bacteria. Peptidoglycan is a vital target because when it is damaged, a bacterial cell explodes. Bacterial cells do not bother to balance the concentration of electrolytes within them with that of their surrounding environment as our cells do. Instead, they let it rise much higher. As a consequence they have an extremely high internal osmotic pressure, or *turgor,* amounting to hundreds of pounds per square inch in most bacteria. That would cause any cell to burst were it not counterbalanced by a strong container. And that is what peptidoglycan does. It forms wall-like chain mail around the bacterial cell, strong enough to withstand its turgor.

In order for the cell to grow, its surrounding wall also has to get larger, and that is a tricky business—enlarging the protective mesh without dangerously weakening it. But it is possible. Imagine a piece of chicken wire. Cutting only one wire at a time does not significantly weaken the mesh. And inserting a new wire unit in the cut strand enlarges the mesh. That is the way a bacterial cell wall enlarges. One enzyme cuts the peptidoglycan mesh and another immediately inserts a larger unit.

Penicillin interferes with this process because its chemical structure mimics that of the new structural unit to be added to the wall. To a certain degree, penicillin even acts like it. Like the new structural unit, penicillin binds to the inserting enzyme. But then the inserting enzyme cannot take the next step. It cannot insert penicillin into the wall. It cannot even get rid of it. So by binding to the inserting enzyme, penicillin inactivates it. All the while the cutting enzyme goes right on cutting more strands in the peptidoglycan wall, quickly weakening it beyond its ability to withstand the bacterial cell's turgor pressure. The cell then explodes.

Penicillin certainly qualifies as a wonder drug, precisely striking a vital target that almost all bacterial cells contain (bacteria from the genera *Mycoplasma* and *Chlamydia* are exceptions) and our cells lack. It kills bacteria without doing us any harm, unless, of course, we become allergic to it after repeated exposure. Small wonder that the impact of penicillin's first use during World War II was so huge. For example, formerly dreaded gonorrhea became so quickly and easily cured that the military had to devise new deterrence schemes for those who contracted this sexually transmitted disease—the "clap shack," for example, where contrite sailors sat on hard wooden chairs during the twenty-four hours of treatment by injection in their gluteus maximus. Penicillin was also highly effective in treating wounds. Those that might previously have led to a lethal infection became readily curable. Almost every bullring in Spain has a street named Calle Alexander Fleming (penicillin's discoverer) nearby. With penicillin, being gored by a bull led merely to a distinguished abdominal scar. Before penicillin, many great Spanish bullfighters were not so lucky, including Joselito in 1920 and Manolete in 1947.

With penicillin widely available, one wonders why any other antibiotics are needed. There are a couple of reasons. First, penicillin cannot reach its target in some bacterial cells because it cannot penetrate their outer protective layers. It cannot penetrate the outer membrane of many gram-negative bacteria, for example, and it cannot penetrate the outer waxy layer of *Mycobacterium tuberculosis,* the bacterium that causes tuberculosis. Penicillin, in its original form, is completely ineffective against these important bacterial pathogens.

Second, natural selection is a powerful force, inevitably yielding new strains of bacteria that are unaffected by penicillin's or any other antibiotic's attack. Because the number of bacterial cells in a population is so huge (several billion in each milliliter is not unusual), natural selection in a microbial setting acts with amazing rapidity. If only one cell in a hundred million or even a billion (and that is about the usual rate of mutation) undergoes a mutation that renders it immune to an antibiotic in the presence of that antibiotic, only it and its progeny will prosper. Soon, within hours under ideal conditions, the entire population of bacteria will be unaffected by (resistant to) that antibiotic.

There are three genetic routes that bacteria can follow to antibiotic resistance: they can exclude the antibiotic from the cell's interior (usually by pumping it out as soon as it enters); they can inactivate the antibiotic by changing it chemically; and they can alter the component within themselves that is the antibiotic's target.

The latter two routes have led to the incidence of penicillin resistance among bacterial pathogens that is now widespread, unfortunately materially diminishing its wonder-drug status. The most rapid and pervasive resistance to penicillin occurred via the first route and affected *Staphylococcus aureus,* the gram-positive bacterial pathogen that causes so many human infections, including the wound infections that killed bullfighters. *S. aureus* quickly acquired the ability to make an enzyme, penicillinase, which rapidly destroys penicillin by opening one of its two constituent chemical rings. When penicillin was introduced into clinical use in the 1940s, essentially all strains of *S. aureus* were exquisitely sensitive and the infections they caused treatable with penicillin. Now almost none are. Penicillinase is most probably a mutationally generated evolutionary progeny of the peptidoglycan-inserting enzyme we just discussed. It retains its ancestral enzyme's capacity to attach to penicillin while gaining the ability to inactivate it by severing one of its rings.

For some reason the genus *Streptococcus,* also containing an important group of bacterial pathogens, did not evolve the ability to acquire resistance to penicillin by making penicillinase. Members of this group remained sensitive to penicillin, and the infections they cause were treatable by the drug for many years. But the power of natural selection is never to be denied forever. Resistant strains of species of *Streptococcus* are now beginning to appear. That is why for the general good some pediatricians are reluctant to prescribe penicillin to treat self-limiting cases of otitis media.

When resistance to penicillin does occur in streptococci, including pneumococcus *(Streptococcus pneumoniae),* it follows the third of the three genetic routes to antibiotic resistance: it mutationally alters the cellular target of the antibiotic. In the case of penicillin, that target is the enzyme that inserts new units into the bacterial cell's peptidoglycan wall. Why should such resistance develop so slowly? Several mutational changes must occur to render an insertion enzyme resistant to high concentrations of penicillin. And bacteria have more than a single kind of insertion enzyme. So a whole package of mutations must occur before a strain of *Streptococcus* becomes sufficiently resistant to penicillin to render the wonder drug clinically useless. Penicillin resistance of streptococci occurs stepwise and slowly, but it does inevitably occur. Resistance happens.

And the story is even bleaker than that. Once a bacterium develops resistance to an antibiotic via the first or third routes (inactivating it or pumping it out of the cell), it can quickly develop the ability to share its newfound skills with other bacteria. The resistance-encoding genes are passed on to plasmids, which can be passed from one bacterium to another by conjugation. Then these conjugative, resistance-encoding plasmids (called R *plasmids*) can spread widely and rapidly to many other bacteria. Some R plasmids encode resistance to several antibiotics. Thus, any antibiotic-resistant bacterial strain has the potential to become a source of widespread resistance. For that reason, unwise clinical use of antibiotics is not the only reason that formerly valuable life-saving antibiotics are now nearly useless in treating many diseases. For example, genes for antibiotic resistance that might develop in innocuous bacteria within a chicken or a hog fed antibiotics to keep it healthy or so it would grow more rapidly could be passed to a plasmid and then to pathogens that might encounter us later on. Antibiotic resistance developed anywhere can be passed on quickly to any other bacterium.

The original penicillin's use in clinical medicine was for the most part limited to treating infections caused by gram-positive bacteria, a group that includes pathogens implicated in conditions as mild as acne and dental caries and as severe as food poisoning, gangrene, pneumonia, scarlet fever, and anthrax, among others. Intense efforts began in the World War II era and later to discover other, more versatile antibiotics. Some of these efforts were directed toward chemically modifying penicillin. They led to a whole set of improved penicillins with names like amoxicillin, dicloxacillin, and ampicillin.

Penicillins are biosynthetically flexible. They consist of a pair of chemical rings and a side chain, which is attached to one of the rings. The rings are the same in all penicillins; the side chains differ. Although the rings are vital to penicillin's activity, the side chains can modulate it, changing, for example, its resistance to stomach acid or the microbes it can penetrate. These variations were discovered in a curious way during World War II while penicillin was being commercially developed. Production began in Britain, where penicillin had been discovered by Alexander Fleming, under the direction of two chemists, Ernst Boris Chain and Howard Walter Florey. All three shared the Nobel Prize in 1945. The side chain of the compound they produced, later designated penicillin F, bore a 5-carbon pentenyl group. This form of penicillin is no longer used. With the exigencies of war, further development became impractical in Britain, and the project was transferred to the U.S. Department of Agriculture's Northern Region Laboratory in Peoria, Illinois. The scientists in Peoria changed the process somewhat. They prepared the growth medium from ingredients that were available locally, one of which was corn steep liquor, a byproduct of wetmilling corn. They were surprised to find that their new process produced a different product. Instead of penicillin F, it yielded penicillin G, which bore a benzyl side chain. Corn steep liquor is rich in the amino acid phenylalanine, which also bears a benzyl side chain. Phenylalanine's side chain had been transferred to the penicillin during its biosynthesis. This flexibility of biosynthesis led to production of a set of biosynthetic penicillins made by adding possible precursors of side chains to the growth medium. One of these, penicillin V, proved useful. It could be administered orally because it was resistant to stomach acid. Then it was soon discovered that penicillin's side chain could be added or modified chemically to produce the larger group of semisynthetic penicillins. These include penicillinase-resistant penicillins, such as methicillin and oxacillin, as well as the broad-spectrum penicillins, such as ampicillin, amoxicillin, and carbenicillin, which are effective against a greater range of hosts, some of which are gram-negative.

Massive screening of microbes for their antibiotic-producing ability also led to rapid successes. The first and probably most dramatic of these was the discovery of streptomycin by Albert Schatz and Selman Waksman in 1943 at Rutgers University. Its properties were spectacular: active against gram-negative bacteria and *Mycobacterium tuberculosis* as well. For the first time, tuberculosis, the dreaded white plague, as it was called, could be effectively treated and cured. But streptomycin did have the major downside of attacking the eighth cranial and thereby frequently causing loss of hearing. Still, many patients were willing to risk loss of hearing to cure their deadly disease.

Penicillin is produced by a fungus *(Penicillium notatum)* that was an accidental contaminant on plates that Alexander Fleming was using to study the wound pathogen, *Staphylococcus aureus,* but streptomycin is made by a bacterium—an actinomycete *(Streptomyces griseus).* Actinomycetes are soil-dwelling microbes, the ones that make soil smell like soil. Because Waksman was a soil microbiologist, very much interested in the actinomycetes that live there, he asked his student, Schatz, to look for antibiotic-producing actinomycetes in soil. And because Schatz was very aware of the cruel impact of tuberculosis, he screened the antibiotic-producing capacity of these actinomycetes directly against *Mycobacterium tuberculosis,* a terribly risky enterprise, because the likelihood of becoming infected was appreciable and curative treatment was not available. Streptomycin was his reward.

With this encouragement an intensive antibiotic search began, largely in the laboratories of drug companies, and it concentrated on actinomycetes from the soil. The successes were rapid and spectacular. Many highly effective and minimally toxic antibiotics were discovered that acted on a variety of targets in the microbial cells—structures that differ significantly from their equivalents in our own cells. As we have seen, the penicillins' target, peptidoglycan, is ideal because our cells lack it completely. But the penicillins aren't the only antibiotics that exploit this vulnerability of bacteria. Other antibiotics, including the cephalosporins and vancomycin, also inhibit the synthesis of bacterial cell walls. The cephalosporins do so the way the penicillins do, but vancomycin takes a different tack. Instead of binding to the peptidoglycan-sealing enzyme, it binds to the new unit of peptidoglycan that the enzyme uses to make the seal. This mechanism of action has proven to be remarkably impervious to the development of bacterial resistance against it. Introduced in 1958, vancomycin remains effective against most gram-positive bacteria, including *S. aureus*.

Other structures in bacterial cells are targets of antibiotics merely because they are different. Chief among these is the bacterial ribosome, which is made up of two components. Streptomycin and tetracycline inactivate the smaller one; erythromycin and chloramphenicol attack the larger. One highly effective antibiotic, refampin, inactivates the bacterial enzyme (RNA polymerase) that transcribes DNA into RNA. Although all cells have an RNA polymerase to transcribe DNA, the one in bacteria differs fundamentally from those found in eukaryotes and archaea. As you would expect, archaea are unaffected by refampin as they are by other antibacterial antibiotics.

Certain antibiotics, including bacitracin, attack uniquely bacterial components of the cell membrane. Because differences between bacterial and eukaryotic membranes are slight, these antibiotics, in general, are restricted to topical use. They are too toxic for systemic use.

A number of synthetic antimicrobial drugs, which are therefore not considered to be antibiotics, also act on targets that are specific to bacteria. The quinolones inhibit DNA gyrase, an enzyme that participates in the bacterial style of replicating DNA. Other extremely useful antibacterial drugs, including trimethoprim and the sulfonamides, act by inhibiting synthesis of the vitamin folic acid. They are selectively active against bacteria because most must make their own folic acid. We do not. We acquire the vitamin in our diets.

With the great success of antibiotic and antimicrobial drugs, euphoria set in. To many it seemed clear that humans were finally victorious in their long struggle against bacterial diseases. Many prestigious specialists in infectious disease declared the battle won. Others remembered Louis Pasteur's famous quotation, *"Messieurs,*

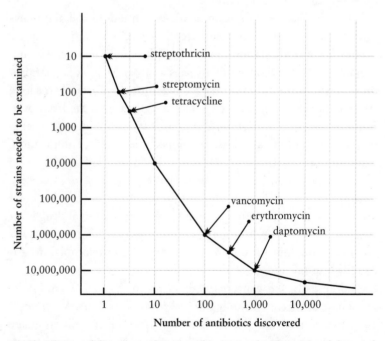

The frequency of discovery of new antibiotics is a log function of the total antibiotics discovered. Several antibiotics are shown with arrows pointing to their frequencies of discovery among random actinomycetes.

c'est les microbes qui auront le dernier mot." ("It's the microbes who will have the last word.")

Very few antibiotics were found that were effective cures for diseases caused by fungi and none for those caused by viruses. Development of bacterial strains that were resistant to effective antibacterial antibiotics was already well under way. A race set in between new discoveries and, a half-step behind, the appearance of bacterial strains resistant to them. The faster they ran, the tougher the race became for the discoverers because new antibiotics became increasingly hard to find. The frequency of discovering a new antibiotic falls (logarithmically) with the number of antibiotics that have already been discovered. Nature's pool of chemotherapeutic riches, quite obviously, is being emptied.

It is only necessary to examine about a hundred soil actinomycetes to find one that produces streptomycin, but as the discovery versus discovered relationship predicts, it is now necessary to examine over 100 million actinomycetes in order to find one that produces a new antibiotic. That is a formidable task. Certainly the long struggle between humans and infectious diseases is far from over. But somewhere along the long evolutionary trail humans, descendants of microbes ourselves, developed the skills of innovation. Those are also formidable. Humans also developed a fascination for making money, but unfortunately the profit motive in seeking new antimicrobial drugs is not that great. The discovery process is increasingly costly, and use of an effective antimicrobial drug is restricted. One takes a drug to treat high blood pressure or other ailments for life, but one course of an effective antimicrobial drug is usually a sufficient cure.

Undoubtedly the concerns, dangers, and impact of antibiotic-resistant pathogens, including *Staphylococcus aureus, Streptococcus pneumoniae,* and *Mycobacterium tuberculosis* will stimulate serious efforts to find new antibiotics or substitute antimicrobial drugs. Methicillin-resistant *Staphylococcus aureus* (MRSA) infections, which used to be restricted to infections acquired while in the hospital, are now spreading through the community. And some of them are deadly. Already more Americans die from MRSA than from AIDS. Of even greater concern is the appearance of strains of MRSA that are resistant to vancomycin, the principal backup drug for treating serious MRSA infections. The first vancomycin-resistant strain appeared in Japan in 1996. Most probably, novel approaches different from the traditional screening of microbes for their ability to produce antibiotics will be explored. One promising approach is similar to one already used successfully in cancer therapy, namely attaching a highly lethal compound to an antibody against the target microbe's surface antigen. The antibody could deliver its lethal partner to the exact location where it should act, thereby opening the possibility of using more widely available classes of toxic agents.

The microbe sighting of the crying child reminds us of the powerful impact of antibiotics on our lives, their now tenuous status, and the dilemma of when to use them and when to do without.

A Stressed CEO Suffers Recurring Abdominal Pains

Until fairly recently, recurring stomach pains among high-level executives were considered almost a badge of honor, the expected side effect of an intense, hard-working personality and a tough job. They offered proof of one's diligence and drive. They should not be taken lightly, however, because peptic ulcers—raw sores in the lining of the stomach or duodenum—can lead to intense pain, hemorrhaging, or even stomach cancer. Before the source of peptic ulcers was fully understood, antidepressants were sometimes prescribed to help sufferers cope and thereby treat their ulcer. Occasionally, doctors blamed spicy food or alcohol as the source of an ulcer.

Diagnosis by endoscopy—swallowing a tube with a camera at the end, which allows the gastroenterologist to examine the stomach and duodenal walls—is definitive. A patient with an ulcer has a distinct sore that is quite apparent. Sometimes, to determine if the ulcer is cancerous, a biopsy is performed using the endoscope to remove a small bit of stomach wall. The treatment, whether or not an ulcer was detected, was usually Tagamet (cimetidine), a histamine blocker that suppresses acid production by the stomach. Sometimes bismuth subsalicylate (Pepto-Bismol), a stomach liner and analgesic, was advised. Usually the patient felt better for a while, only to return in six months or so for a similar treatment. Some ulcer patients were subjected to a bland-food diet, which they endured for life.

In the early 1980s, J. Robin Warren, an Australian pathologist, noticed unusual-looking bacterial cells in many of the biopsy tissues he had taken from patients with ulcers. He suggested to his gastroenterologist colleague, Barry Marshall, that this apparent correlation between bacterial cells and ulcers was worthy of investigation. Indeed, the correlation held up and Warren was eventually able to cultivate in the laboratory these bacteria, which were named *Helicobacter pylori*. But correlation does not prove causation. Even the fact that antibiotics that killed *H. pylori* cured his patients' ulcers was not convincing to leaders in the field of ulcer study and treatment.

Robert Koch, a German microbiologist and physician, addressed the correlation/causation dilemma about a hundred years previously. During the 1880s he was engaged in research to determine if *Mycobacterium tuberculosis* causes tuberculosis. Not until he caused healthy experimental animals to develop tuberculosis by administering *M. tuberculosis* cells were he and other scientists convinced of a causal connection between *M. tuberculosis* and the disease. But Warren was unable to find a suitable experimental animal for his studies on *H. pylori* and ulcers. The pigs he tested suffered no ill effects when given *H. pylori*.

Marshall's interest, of course, was whether *H. pylori* causes ulcers in humans. Why not try that? So he drank a suspension of about a billion *H. pylori* cells. On the fifth and sixth days after imbibing, he vomited in the morning and developed bad breath. After a week his stomach felt like a "lump of lead." On the eighth day an endoscopy revealed the beginnings of an ulcer with *H. pylori* cells attached to it. On the fourteenth day he confessed to his wife what he had done. She insisted that he take antibiotics, which he did. The experiment was over.

In 1984 Marshall published an account of his dramatic experiment in the *Australian Medical Journal,* convincing most scientists that *H. pylori* can cause peptic ulcers. Along with J. Robin Warren, he was awarded the Nobel Prize in Physiology or Medicine in 2005 for the discovery. By then, medical opinion had changed fundamentally: stressful living does not cause ulcers (although it could exacerbate them). A distinctive, helically shaped microbe does. And administering antibiotics can cure them.

Infection and consequence, however, are not as immediate as they are in many other diseases. Sometimes years separate infection and first ulcer symptoms, such as pain or vomiting. Excessive production of stomach acid must contribute to the development of stomach lesions. And other factors, probably including stress, certainly contribute. Also, the connection between *H. pylori* infection and having ulcers is not absolute. Approximately 30 percent of Americans are infected, but only about 10 percent develop peptic ulcers. When they do, most can be effectively treated with antibiotics that eliminate *H. pylori*. Worldwide, it is estimated that 50 percent of humans are infected by *H. pylori,* making half the world's population vulnerable to peptic ulcers. *H. pylori* infection increases the odds of more serious diseases such as stomach cancer and possibly cholera, and so it presents a real public health problem. Some believe that a vaccine would be the best way to attack it. Triple drug treatment (two antibiotics—metronidiazole, tetracycline, calithromycin, or amoxicillin—plus either an acid blocker or stomach-liner shield) for two weeks cures 90 percent of patients. But it is expensive, possibly unfeasibly so for worldwide use. Curiously, Pepto Bismol's efficacy against peptic ulcers is now comprehensible: in addition to offering symptomatic relief, it kills *H. py-lori.*

A microbial etiology of peptic ulcers always seemed improbable. How could a microbe survive and prosper in the acidity of the human stomach, an environment with a pH almost as low as battery acid? *H. pylori* does so by making itself a hospitable microenvironment there: it produces an enzyme, urease. As we learned when we examined the manure pile, urease splits urea, forming carbon dioxide and ammonia. Ammonia is a base. By making it from urea, *H. pylori* neutralizes the stomach acid in its immediate surroundings.

One novel way to detect an *H. pylori* infection is to administer a breath test, which is accurate in 96 to 98 percent of all cases. The test's accuracy relies on the presence of urease in the patient's stomach. The patient is given a small dose of urea labeled with radioactive (^{14}C) carbon. If he or she is infected with *H. pylori,* the bacteria's urease will split the administered urea into ammonia and radioactive carbon dioxide, which can be detected in the patient's breath.

H. pylori is now established as the most common cause of peptic ulcers, but it is not the only one. Excessive use of pain relievers such as aspirin and ibuprofen can bring one on, and factors including smoking, excessive alcohol consumption, and, indeed, stress may certainly contribute. But with reasonable assurance the busy CEO and many others with ulcerlike distress play unwitting host to a microbe.

The Dentist's Drill

If the dentist is brandishing his drill to attack a dental caries (a Latin substitute, both singular and plural, for the less graceful English word "rot"), his whirling drill that none of us relishes is most probably, although indirectly, the result of a sighting of one particular microbe, the bacterium *Streptococcus mutans.* Although this lactic acid bacterium is not the only one capable of making holes in our teeth, it is by far the major culprit. Dental caries can be viewed as a specific, chronic, but treatable bacterial infection caused by *S. mutans*, which earned its curious name when it was first cultivated in 1924. Those that studied it considered it to be a mutant strain because its cells are more oval than round, as the cells of most streptococci are.

Different strains of *S. mutans* have certain properties that are sufficiently variable to encourage some dental microbiologists to subdivide the single species into several or refer to it as the *S. mutans* group. Nevertheless, all strains of *S. mutans* share a core set of characteristics that seem perfectly designed for this bacterium to make holes in our teeth, especially if we consume table sugar, or sucrose. Being a lactic acid bacterium, *S. mutans* acquires metabolic energy and makes its PMs by fermenting sugars. And like all microbes that depend exclusively on fermentation, *S. mutans* produces large amounts of the products of fermentation in order to

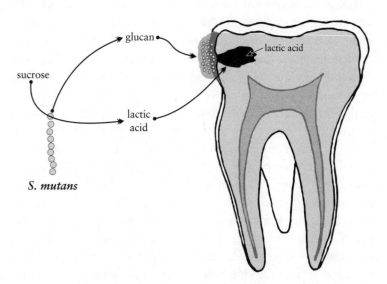

Streptococcus mutans makes dental caries.

obtain adequate amounts of metabolic energy for growth. The fermentation product that *S. mutans* and similar bacteria produce, lactic acid, is a strong acid, capable of dissolving the surface enamel that protects our teeth, especially if the acid is produced right on the surface of the tooth. And that is what *S. mutans* does when it attaches firmly to a tooth.

Lactic acid softens the tooth's surface (demineralizes it). If the exposure to lactic acid is brief, components of saliva repair the damage completely (remineralize it). Moreover, saliva, because it is slightly alkaline, can neutralize small amounts of lactic acid. So saliva is a natural and highly effective caries fighter. The rampant, tooth-destroying outbreak of caries that users of methamphetamine experience is tragic witness to saliva's importance in defending against dental caries. Methamphetamine causes "dry mouth" by inhibiting the formation of saliva, leaving users' teeth exposed to higher concentration of lactic acid and unable to repair its initial damage.

If the lactic acid is not neutralized and exposure is prolonged, the softened region of the enamel disintegrates, opening a hole through which lactic acid enters and attacks the more vulnerable underlying dentine. By destroying dentine, lactic acid produces the cavity that the dentist is planning to clean and shape with his drill. If the dentist does not afterward plug the cavity with silver amalgam, plastic, gold, or some other suitable material to shield the tooth's exposed interior from further attack by lactic acid, it will continue to enlarge, perhaps entering the root canal and descending through it to underlying tissues, inviting the formation of an abscess (a pocket of bacterial infection). Then more desperate and heroic dental intervention—a root canal or an extraction—is required.

In addition to being able to make caries-forming lactic acid from any of a number of sugars, *S. mutans* can make another substance, a gummy polysaccharide called *glucan* (a large number of glucose molecules strung together forming a chain), which confers to *S. mutans* its unique caries-forming potential. However, *S. mutans* can make glucan from only a single sugar, sucrose. It does so by synthesizing and excreting an enzyme, glucosyltransferase, that splits the disaccharide, sucrose, into its component sugars, glucose and fructose. As the enzyme severs sucrose, it links together the glucose halves of the sugar to grow the molecule of glucan, releasing the fructose half to be fermented into lactic acid by the *S. mutans* cell. Glucan augments caries formation in two ways. First, it sticks to the surface of a tooth so that *S. mutans* cells may attach to the tooth as well (via specific glucan-binding proteins on its cell surface). Second, glucan forms a thick barrier that excludes saliva with its neutralizing and enamel-restorative capabilities from access to the tooth's surface.

Once *S. mutans* is ensconced on a tooth, any sugar, fructose from fruit, lactose in milk products, or glucose from a variety of sources, can continue caries formation as *S. mutans* ferments them to lactic acid. Even starch in cereal grains contributes because an enzyme (amylase) in saliva converts it to glucose, which *S. mutans*, in turn, converts to lactic acid. In other words, most of the foods we consume contribute to caries formation once *S. mutans* is established. Each time we consume one of them, a burst of lactic acid formation lasting an hour or so occurs. Unsurprisingly, there is a strong correlation between frequency of eating and formation of caries. Snacking, it seems, affects more than our waistlines.

The archeological record tells a convincing story about diet and caries. It is, indeed, an ancient disease. Teeth in million-year-old skulls show clear signs of caries. The frequency of caries increased during the Neolithic period, when plant foods containing carbohydrates became a greater component of human diet. The history of Native Americans tells a similar story. Their incidence of caries increased with the arrival of European colonists, a period when Native Americans started eating more corn.

But in spite of the relative increase in caries brought about by consuming cereal grains, the incidence of caries remained low during the Bronze Age. Prevalence jumped dramatically, however, in about 1000, when sugarcane, a particularly rich source of sucrose, became available in Europe. Most plant products, even particularly sweet ones, contain only small amounts of sucrose. Honey, for example, which is about 80 percent sugar, contains just over one percent sucrose, and the berries of grapes do not contain any. In contrast, almost all the sugar in sugarcane, as well as sugar beets, is sucrose. Now, with the spread of sugary western food, including infant formula, rotting teeth are epidemic in much of the world. In China, 75 percent of five-year-olds now have tooth decay.

Proof of the central role of *S. mutans* in causing caries rests on more than plausibility. Studies on experimental animals, principally rats, firmly establish the connection between *S. mutans* and caries. Most of us become infected with *S. mutans* at an early age, most frequently, it is presumed, from a parent's kiss. Of course, the obvious solution to caries would seem to be preventing *S. mutans* from reaching your mouth or else eliminating it once it is there. Such approaches to control are under way, some with impressive results, in experimental animals. One intriguing and promising approach is analogous to protecting plants from freeze injury by displacing their normal population of ice-nucleating bacteria with similar bacteria that occupy the same niche on the plant's surface but do not nucleate ice. Another example of this probiotic approach is used to protect baby chicks from *Salmonella* infections.

J. D. Hillman at the University of Florida, who developed such a *S. mutans*–displacing microbe, calls the approach "replacement therapy." By eliminating certain genes from a clinical isolate of *S. mutans,* he developed a strain that grows well but does not produce lactic acid and can crowd out the mouth's normal population of *S. mutans* by occupying its particular niche on the surfaces of teeth. But Hillman went one step beyond crowding to destroying and occupying. He introduced into the replacement strain the capacity to make a highly specific, antibioticlike protein that kills other strains of *S. mutans.* Such highly specific killer proteins, called *lantibiotics,* are widespread in the world of gram-positive bacteria. Like the one produced by Hillman's strain, they are restricted to internecine microbial warfare. Some have been used for over thirty years to preserve dairy products.

Trials of Hillman's replacement therapy for caries on rats have been spectacularly successful. Given to young rats, the replacement strain of S. *mutans* provides lifelong protection, virtually eliminating caries from rats fed a high-sucrose, caries-promoting diet. Human trials are under way. The treatment, termed SmaRT for "S. *mutans* and replacement therapy," would only involve a dentist's squirting a suspension of the altered S. *mutans* onto the tooth surfaces. Then after about five minutes, the patient would be advised to go home and eat something sugary to help the genetically modified bacterium become firmly attached to the patient's teeth. Perhaps it would confer lifelong protection against caries for humans as it has for rats.

Another approach to eradication of caries, a worldwide goal that the global task force of chief dental officers has set for children during the next twenty-five years, is vaccination. The target of most vaccines now being evaluated is the glucan-forming enzyme glucosyltransferase, which S. *mutans* secretes. So the vaccine, which would be administered as an oral spray designed to stimulate production of antibodies in saliva, would be preventative and given at an early age, before an infant becomes infected with S. *mutans*.

Fluoridation, administering fluoride ion (F^-) to the surface of teeth by way of the water supply, toothpaste, mouthwash, or dental application, is a well-known approach to controlling caries. Rather than controlling S. *mutans*, it depends on making the tooth more resistant to attack by hardening the tooth's enamel surface to avoid lactic acid's destructive powers. Administered fluoride replaces hydroxyl groups ($-OH$) in the hydroxyapatite of the tooth's enamel, converting it to much harder and more lactic acid–resistant fluoroapatite. Studies conducted before the widespread administration of fluoride by other means show that fluoridation of municipal water supplies at the minuscule levels of 0.7 to 1.2 parts per million decreases the incidence of caries by 40 to 60 percent at a cost of only a few cents per inhabitant per year. As of 2002, 67 percent of Americans were living in communities supplied with fluoridated water. Still, fluoridation of municipal water remains highly controversial, possibly because considerably higher levels of fluoride cause mottling of developing teeth, or because it is the active ingredient in certain rat poisons, or as a result of generalized suspicion of government intervention. If the goals of the global task force of chief dental officers are met, perhaps one day in the not-too-distant future we will be able to cross this microbe sighting off our list.

A Bottle of Apple Juice, Vintage 1996

Escherichia coli, that otherwise benign laboratory workhorse, can turn nasty. *E. coli* O157:H7 is best known for causing bloody diarrhea. And the infection has the potential to develop into a more serious disease, hemolytic uremic syndrome (HUS), which can lead to kidney failure, other complications, and death, particularly among infants and the aged.

The mysterious numbers and letters that often follow the name *E. coli* distinguish a particular strain among the myriad other strains of *E. coli,* a bacterium that is, by now, quite familiar to experienced microbe watchers. The various strains of *E. coli* are distinguished traditionally by quite minor differences that occur in the form of certain proteins on the outsides of their cells. Even though variations among these proteins are minute, they can be detected by taking advantage of the extraordinary discriminatory power of antibodies. One simply determines whether or not a particular strain reacts with antibodies prepared against other isolates of that same strain. Because antibodies are used to detect them, such cell surface–exposed proteins are referred to as *antigens* and the distinguished strains as *serotypes.* The letter "H" is the initial of the German word *hauch* ("breath"), which refers to a protein in the cell's flagella; "O" (*ohne*, "without," because it is detectable in cells that lack flagella) refers to a polysaccharide on the cell's outer surface. So the culprit strain gets its designation because it caries the seventh serologically detected form of the flagellar protein and the 157th of the outer surface polysaccharide.

E. *coli* attained its unmatched popularity as a focus organism for genetic and biochemical studies because of its reputation for being easy and safe to work with and having a highly flexible metabolism. In the presence of air, it acquires metabolic energy by aerobic respiration. In its absence, E. *coli* can shift its metabolism to acquiring energy by one of the several types of anaerobic respiration or, if oxygen substitutes are also unavailable, by fermentation. E. *coli* can do it all. It can acquire metabolic energy by all known routes with the exception of photosynthesis. When scientists realized the remarkable unity of the fundamental aspects of biology—metabolism, biochemistry, molecular composition, and genetics—they were encouraged to study the most readily accessible organism. E. *coli* became the favorite. Jacques Monod, a Nobel Prize–winning student of the way E. *coli* expresses its genes, famously but as we know now a little too expansively declared, "Anything that is true of E. *coli* must be true of elephants, except more so."

The focus on E. *coli* became autocatalytic: as more was learned about it, it became easier to study, and more could be learned more easily and quickly. The collective scientific enterprise of concentrated study of E. *coli* turned out to be richly rewarding. An astoundingly large portion of our fundamental knowledge about general biology has come from intense study of this somewhat obscure, minority component of our intestinal microbiota as well as those of other animals, including cold-blooded ones. EcoCyc.org is a Web site well worth visiting. It gives a glimpse of the wealth and completeness of information now known about this bacterium. The function of almost all of its 4,576 genes is known. In contrast, we are not even sure how many genes humans have.

Because so much is known about E. *coli,* it became a near indispensable tool of biotechnology. Genes to be manipulated and studied for whatever reason are almost always first cloned into E. *coli.* And a number of therapeutic human proteins, including human growth hormone and insulin, as well as certain antibodies, are manufactured in large fermenters by strains of E. *coli* into which the encoding human genes have been introduced.

Why E. *coli?* It is an often-asked but tough question. As we noted, it is versatile and amenable to study, but certainly not uniquely so. Possibly, E. *coli* got its first push toward stardom simply by being an easy-to-identify inhabitant of our intestines: it can ferment lactose (not many bacteria can); it can grow at 45°C (113°F) (not many microbes that live in our intestines can); it grows into easily recognized, metallic-black colonies on an easily prepared medium (eosin methylene blue, EMB agar); and its colonies produce considerable quantities of readily detected acid from sugars. As such, E. *coli* was deemed useful and appropriate as a test of whether a water supply had been contaminated with sewage. The rationale is uncomplicated. E. *coli* is an inhabitant of our intestines. If it is found in a water supply, it probably got there through sewage. Freedom from E. *coli* became the legally recognized index of microbiologically safe water. All university departments of bacteriology kept cultures of E. *coli* so they could instruct their students how to test whether water is safe.

The next big step for E. *coli*'s rise to fame occurred in the early 1950s when a young medical student, Joshua Lederberg, at the University of Wisconsin at Madison wanted to test what then seemed a preposterous question. Do bacteria exchange genes? Do they mate? Because cultures of E. *coli* were available in the bacteriology department's culture collection, he chose them to experiment with. Indeed, he found these cultures did exchange genes, and they did so by conjugational mating. This momentous discovery created the opportunity to make genetic crosses between strains of E. *coli* and thereby to map the location of genes on its chromosome and then to study the composition of individual genes and their function in the cell. Lederberg's discovery of the sexual behavior of *E. coli* firmly established that microbe as the premier organism of experimental biology. It was then well on its way to attaining its present status of being the most thoroughly understood of all cellular organisms.

But *E. coli* always did have its dark side. Although the vast majority of strains of *E. coli* are completely harmless and might even be beneficial in protecting us from pathogenic bacteria such as *Shigella* and *Salmonella* by occupying their preferred sites on the intestinal wall, some strains of *E. coli* are pathogens. These strains acquired plasmids that encode *virulence factors* (properties that confer the ability to cause infections and disease).

The most common disease caused by *E. coli* is a urinary tract infection. And *E. coli* from the intestinal tract is the most common source. Strains of *E. coli* cause 90 percent of urinary tract infections in otherwise healthy people. Strains of *E. coli* that cause urinary tract infections present plasmid-encoded pili on their outer surface that bind specifically to receptor proteins on the surface of the bladder, allowing them to cling tightly enough not to be washed out by the flow of urine. Women are the most frequent sufferers of urinary tract infections because they have shorter urethras. Some women who experience frequent recurrent infections tend to have larger numbers of *E. coli* receptors in their urinary tracts.

Plasmid-bearing strains of *E. coli* also cause various diarrheas, including those known as traveler's diarrhea and infant diarrhea. The latter is a major cause of infant suffering and death in developing countries. In addition to adherence factors, diarrhea-causing strains produce one or another of two plasmid-encoded toxins, called *heat-stable* and *heat-labile* toxins.

But *E. coli* O157:H7, which burst on the scene in 1982 causing an outbreak of gastrointestinal disease spread by contaminated hamburger, is quite different from other known strains of *E. coli*. It does not merely carry an added plasmid as previously known pathogenic strains of *E. coli* do. It has a fundamentally different set of genes. Its chromosome is much larger, with almost a thousand genes not found in other strains of the bacterium. The chromosome of *E. coli* K-12, the most commonly used laboratory strain, encodes 4,576 genes; *E. coli* O157:H7's encodes 5,476 genes. In addition to having more, *E. coli* O157:H7 also has less. It cannot grow at 45°C, and it is unable to utilize the sugar alcohol sorbitol as a nutrient; both abilities are defining characteristics of most other strains of *E. coli*. The dangerous characteristics that *E. coli* O157:H7 strains had acquired are disease-causing toxins called *shigalike toxins*. Such toxins can severely damage the lining of the intestine, causing diarrhea to be bloody. *E. coli* O157:H7 probably got the encoding genes as a consequence of their being transferred from *Shigella,* a long-standing bacterial pathogen that causes severe gastrointestinal disease.

The source of the *E. coli* O157:H7 that caused the 1982 hamburger outbreak of gastrointestinal disease was traced to an infected cow, and the extent of the outbreak was a consequence of the way hamburger is distributed commercially: it is prepared in large batches, which are subdivided for wide distribution. It soon became clear that cattle carry this strain of *E. coli*. In some herds the majority does. The bacterium does not bother them, but it is passed in their feces. It seemed clear that the hamburger must have become contaminated with cow manure. In succeeding years a number of other outbreaks were attributed to *E. coli* O157:H7. For the most part, these, too, were traced to contaminated hamburger and in one case, salami.

In 1996 an outbreak of *E. coli* O157:H7 in bottled apple juice introduced a new route to infection. Apparently, cattle had grazed in the apple orchard and fallen apples gathered to make juice were contaminated. The juice was not pasteurized; the process was deemed not natural. The outbreak led to several dozen young children developing hemolytic-uremic syndrome and to the death of one of them. Hemolytic-uremic syndrome can result in permanent loss of kidney function. In the United States this disease is now the principal cause of acute kidney failure in children.

Since then, E. coli O157:H7 infection has come to be associated with fresh vegetables, including bean sprouts, tomatoes, spinach, and other packaged greens. The bacterium clings tenaciously to the surfaces of vegetables and cannot be removed by washing. Somehow the bacterial cells are transferred from carrier animals to the vegetables. Ten such outbreaks were reported between 2002 and 2006. E. coli O157:H7 is also transmitted in milk and can be acquired by swimming in polluted water.

In 2006 a nationwide outbreak was traced to bagged baby spinach leaves from the central California coast. This was a major outbreak: more than 200 people spread over twenty-six states became ill. What was its source? The fields where the infected spinach was grown had been carefully fenced to exclude cattle. The suspected animal reservoir for the outbreak is feral pigs, which are as common in that part of California as they are in many other places. Feral pigs easily burrow holes under the protective fences to gain access to the spinach, which they appreciate. Of course, other animals such as small rodents or birds also could be culprits.

Once an outbreak is started, it can spread from person to person, particularly among the young in daycare facilities and the elderly in nursing homes. It is hard to stop the cycle because the bacterium, like Shigella spp., is so highly infective. Stools of an infected patient contain huge numbers of bacterial cells and as few as ten of them are thought to be able to transmit the disease.

Most people who contract an E. coli O157:H7 infection suffer minor symptoms and recover uneventfully. So many cases go unreported. In spite of the number of missed cases, E. coli O157:H7 is estimated to be second only to Salmonella as the cause of bacterial diarrhea in the Pacific Northwest. And possibly there are as many as 75,000 cases annually in the United States, many of which require hospitalization. E. coli O157:H7 does much more than besmirch the good name of E. coli.

Treatment of the infection is only palliative: patients should be kept hydrated and comfortable. Antibiotic therapy is not recommended. Although *E. coli* O157:H7 is susceptible to many antibiotics, the dying bacterial cells release more toxin, worsening the patient's condition. Antidiarrheal medicines such as Imodium are also to be avoided, on the basis of similar reasoning.

The bottled apple juice containing *E. coli* O157:H7 brought to light a new microbe, one that probably did not exist forty or so years ago. It reminds us how quickly new microbes can evolve and how rapidly and extensively modern human lifestyles can spread them. The world of microbes is as constantly changing as our own. *E. coli* O157:H7 probably evolved from an ordinary, innocuous strain of *E. coli* when it received chunks of genes from other bacteria, including, most probably, a species of *Shigella,* another bacterial pathogen. Analysis of *E. coli* O157:H7's DNA strongly suggests that a bacterial virus (a phage) mediated the transfer, that is, that the transfer occurred by transduction.

Microbes' roles as pathogens are familiar to all of us. They affect both our lifestyle and our life span. But their effect on Earth's weather and even its geology may be less familiar to us. We will encounter some of their planetwide impacts in the next chapter.

12

Shapers of the Planet

Tiny as they are, some microbes do very big things. They shape our weather, Earth's geology, and its environment. In this chapter, we will witness some of these big things.

A Snowmaking Machine

A machine spewing a dense plume of snow on a thinly covered downhill run is a welcome sight for skiers and an esthetically spectacular display, particularly on a windy day. It rivals fireworks. Water and compressed air are pumped out of the gun, creating a cascade of small droplets. Under most conditions, each droplet deliberately contains one cell of the bacterium *Pseudomonas syringae*. The bacterial cells, more properly bacterial corpses, are added as a slurry made from dry powder to the water being pumped into the gun.

The powder, a preparation of dried, irradiation-killed bacterial cells, is made commercially and sold in approximately one-pound pouches under such trade names as Snomax. A pouchful is sufficient to treat at least 100,000 gallons of water. These bacteria con-

tain on their outer surface a particular protein with a powerful
ability to act as a nucleating agent, that is, to serve as a focus upon
which ice crystals can form and grow to become a snowflake. This
nucleating protein is extremely rare in nature. Only *P. syringae* and
a few close relatives, less than about half a dozen species, are known
to make it.

The story of the discovery of these microbial needles in the biological haystack provides an interesting look at how divergent lines
of scientific investigation sometimes converge. The discovery was
made about thirty years ago simultaneously and independently by
two groups of microbiologists, one in Wyoming and the other in
Wisconsin, stimulated by quite different observations.

The Wyoming group was intrigued by a puzzling meteorological
phenomenon. They and others were aware that in their semiarid
region where rainfall is especially prized and carefully monitored,
vegetation-covered areas receive greater rainfall than more barren
ones. Of course, there is an immediate question of cause and effect. Does the rainfall stimulate plant cover or is it the other way
around? The meteorologists' hypothesis—that something on the
leaves of plants might be swept into storm clouds by prestorm
winds, thus seeding the clouds and acting as nucleating agents, as
silver iodide was known to do—proved to be correct. The microbiologists then discovered that the leafborne agent was *Pseudomonas
syringae,* a ubiquitous leaf-dwelling bacterium that sometimes
causes plant disease. Such nucleating agents cause water vapor to
coalesce into raindrops as well as to serve as foci around which ice
crystals form.

The microbiologists in Wisconsin followed a quite different guiding observation, namely that early spring frost damage to seedling
corn became more severe if the corn seedlings had been dusted with
plant debris. Whether the dusting occurred inadvertently while till-
ing or intentionally in the Wisconsin microbiologists' tests made
no difference—the seedlings suffered the same outcome. Sorting
through the debris revealed that the leafborne bacterium *Pseudo-*

monas syringae was again the active agent. Here, rather than causing water vapor to coalesce into droplets, it had served as a nucleating agent for ice crystals to form on the corn seedlings, causing frost damage.

As the term *frost damage* suggests, low temperature in and of itself does not damage a plant; ice crystals do. Their formation punctures membranes, ripping apart plant cells and killing them. In the absence of a nucleating agent, ice does not necessarily form at the freezing point of water, 0°C (32°F) (a phenomenon called *supercooling*), although, of course, it does melt there. Rigorously pure water that lacks any nucleating agent, even dust (which has low-level nucleation ability), can be chilled to as low as −40°C (−40°F) without freezing. So spring temperatures somewhat below the freezing point do not bother the corn seedlings because they remain frost-free. Those seedlings that have been dusted with plant debris, however, are damaged because the *P. syringae* it contained nucleated the formation of cell-destroying ice crystals.

Below about −5°C (23°F), frost damage occurs even in the absence of *P. syringae* because other, less effective nucleating agents, including some produced by the plant itself, prevent further supercooling. But *P. syringae* cells, which are almost ubiquitous on leaves, are responsible for the ice formation and frost damage to plants that occurs in the range of 0 to −5°C.

A series of observations established unequivocally that the nucleation activity of *P. syringae* resided totally in one particular protein, InaX (for ice nucleation activity), on the bacterium's outer surface. The most powerful indictment of InaX was the observation that mutant strains that lack InaX, known as *ice-minus*, as well as species of bacteria that were closely related to *P. syringae* but also lacked InaX all were without nucleation activity. Even more convincingly, perhaps, when the gene *(inaX)* encoding InaX is transferred to *Escherichia coli*, that normally nucleation-incapable bacterium gains the ability to nucleate. Without doubt, the nucleation capacity of a bacterium depends on its ability to make InaX.

P. *syringae*'s ability to nucleate ice formation has been put to practical use in snowmaking machines as described, permitting them to operate effectively at temperatures as high as minus a half a degree Celsius (31°F), even if the humidity is low. Harnessing InaX's rainmaking ability by seeding clouds in order to augment local rainfall has only been considered. But ice-minus mutant strains of P. *syringae* have been used successfully to do the opposite of snowmaking—to prevent ice formation and thereby protect plants from frost damage. Normally occurring, frost-inducing strains of P. *syringae* on the leaves of plants are replaced with an ice-minus strain (one genetically unable to make InaX). Even partial replacement of the normal microbiota is beneficial because the amount of frost damage to a plant's leaf that occurs slightly below the freezing point is directly proportional to the number of ice-plus P. *syringae* cells (those able to make InaX) on that leaf. Spraying plants with an ice-minus strain, which competes for the same ecological niche of the plant surface, diminishes the number of ice-plus cells that are present, and thereby extends the frost-free temperature range downward by several degrees, a very important advantage for many crops. The ice-minus strain does not outcompete the ice-plus strain. It merely occupies an area of the leaf surface, preventing the ice-plus strain from establishing itself there.

There can be no doubt that P. *syringae*'s activity as a powerful nucleating agent rests exclusively on its formation of one particular protein on its outer surface. No other activity of or function for the protein has been discovered. As we noted, mutant strains (ice-minus) that cannot make this protein are seemingly not diminished in other respects: they proliferate and occupy space on the surface of a plant leaf as successfully as their parental (ice-plus) strains do. Why, then, would the ability to make such a highly specialized protein as InaX have evolved? Perhaps it offers the selective advantage of exploiting low temperature to rupture plant cells, thereby enriching P. *syringae*'s meager supply of nutrients on the leaf surface with material exuding from the leaf cell's much richer interior.

In field trials conducted in the early 1990s, spraying young potato plants with an ice-minus strain worked wonderfully well. It decreased the population of wild-type ice-plus strains fiftyfold and reduced damage from a subsequent overnight frost by 80 percent. But in spite of its effectiveness, the procedure has not found commercial success, largely because the ice-minus strain is obtained using methods of recombinant DNA technology. Although genetically equivalent and probably genetically identical strains appear regularly in the environment as a consequence of inevitable natural events, regulatory approval of this new technology required the applicants to conduct field tests that took over five years. No untoward consequences were detected, yet public concern over releasing recombinant DNA-generated microbes in nature stopped commercialization of the method. Curiously, the phrase "genetically modified" inflames intense public passion, in spite of humankind's utter dependence on crops that were highly genetically modified from their original naturally occurring stock centuries ago. The wild ancestors of our major cereal crops—rice, wheat, and corn—were puny plants that could not begin to support today's burgeoning human population. Teosinte, corn's forbearer, for example, hardly resembles its modern descendant. All these civilization-saving accomplishments were genetic modifications.

The studies on ice-minus bacteria were not in vain, however. They established the feasibility of preventing frost damage by displacing a plant's resident population of *Pseudomonas syringae*. Soon it was discovered that a naturally occurring strain of a closely related bacterium, *Pseudomonas fluorescens*, displaces ice-plus strains of *P. syringae* almost as effectively as ice-minus strains of *P. syringae* do. The use of *P. fluorescens* for this purpose has become widespread. In 2004, about 50,000 pounds of freeze-dried cells of *P. fluorescens* were sold to prevent frost damage to plants, probably the largest use of any biological control agent.

Using a benign microbe to displace harmful microbes is hardly unprecedented. As we discussed in an earlier chapter, the technique is being investigated for treatment of dental caries. Also, the U.S. Department of Agriculture developed a probiotic method to control infection of baby chicks by *Salmonella.* Administering to chicks a mixture of the bacteria found in the guts of normal healthy young chickens decreases the incidence of *Salmonella* infection from 90 percent to 10 percent. Presumably this simple, effective treatment works in much the same way as frost-control treatment. The benign microbes occupy the available environmental niche, in this case the chick intestine, making it unavailable for the undesirable one, in this case a potentially deadly pathogen.

Whether naturally occurring, like *P. fluorescens,* or engineered by humans, like ice-minus strains of *P. syringae,* microbes have evolved a variety of strategies to compete in nature, and many of these strategies intentionally or unintentionally impact our lives and the environment, even influencing our daily weather.

A Fluffy White Cloud over the Ocean

Even inveterate microbe watchers might be surprised to learn that microbes in the ocean cause clouds to form above it. But increasing evidence shows they do.

Certain eukaryotic microbes (protists), principally dinoflagellates and their relatives that make up the ocean's phytoplankton, instigate a cascade of events that ends with cloud formation over the ocean. We have already encountered dinoflagellates, those small points of light in the wake of a ship. Dinoflagellates are a distinctive group of unicellular, photosynthetic protists that constitute a major faction of the ocean's phytoplankton. They are readily identifiable microscopically by their distinctive appearance and their curious twisting, spiral motion.

For protection against the destructive effects of solar radiation, these floating microbes produce a sulfur-containing substance called dimethylsulfoniopropionate (DMSP). DMSP remains inside young cells that are growing vigorously and keeps them intact. If solar radiation becomes more intense and water temperature rises, requiring more protection, the cells respond by accumulating even more DMSP. But when the cells begin to grow old and senesce, they convert some of their DMSP into dimethyl sulfide (DMS), and both DMSP and DMS leak out of them. Surrounding bacteria take up the DMSP and convert it to DMS as well. So eventually all the DMSP that the dinoflagellates have produced ends up as DMS. And unlike DMSP, DMS is volatile and enters the atmosphere. Altogether huge quantities (estimated to be over 50 million tons per year) of microbially generated DMS rise from the oceans into the atmosphere. There, solar radiation converts DMS into submicron-sized granules of sulfate, forming an aerosol that nucleates water vapor and causing clouds to form. And once clouds have formed, the aerosol stabilizes them by slowing their aggregation into water droplets. So microbes cause clouds to form and persist longer.

Clouds impact our climate by cooling the earth. They shade its surface and they reflect solar radiation back out of the atmosphere.

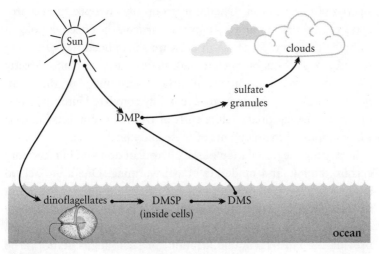

Microbes making clouds.

Increased cloud formation would certainly moderate but not arrest the impact of global warming. And evidence suggests that will be the case. Warming of the oceans is causing the blooms of dinoflagellates and related protists to become more abundant and to move farther north—yet another example of microbes' tendency to maintain Earth's ecology in life-sustaining balance. Notice that microbe-mediated cloud formation has a self-adjusting, built-in feedback loop. As intensity of sunshine increases, dinoflagellate stress increases; they produce greater quantities of DMSP, which forms more clouds. That cloud formation decreases intensity of sunshine and attendant dinoflagellate stress, causing them to make less DMSP, and so on.

Phytoplankton, which occupy the sunlit layer of the ocean that extends down some 600 feet from the surface, are a major force for other environmentally balancing acts as well. They supply 50 to 90 percent of the oxygen in the atmosphere, depending on seasons. And collectively, they are the world's principal users of its carbon dioxide. Billions of tons of atmospheric CO_2 are sequestered on the ocean floor in carbon-containing material that is the residue of sunken phytoplankton. One large group of phytoplanktonic protists, the coccolithophores, plays a particularly effective role in trapping carbon dioxide. Their cells are surrounded by beautifully sculptured platelets of calcium carbonate ($CaCO_3$), made directly from CO_2. After these outer cell surfaces accumulate on the ocean floor in huge deposits, they are eventually converted into limestone or chalk. The legendary white cliffs of Dover in southern England are examples of the remnants of these microbes.

Increasing the population of the phytoplankton would trap more carbon dioxide and moderate global warming. One approach to augmenting their abundance consists of adding iron to the ocean because growth of phytoplankton in the ocean is limited by the availability of this essential nutrient. Probably the southern oceans provide the most favorable location for such augmentation. The southern oceans are rich in the other essential nutrients that phytoplankton need, notably nitrogen. Experimental trials have established that added iron does indeed increase the density of phytoplankton. Such trapping is only transient, however, unless the phytoplankton sink to the ocean floor. Otherwise, as the phytoplankton die they are metabolized by bacteria, which reconvert their carbon content into CO_2, which is then released back into the atmosphere. In the experimental trials, the fraction (only a few percent) of the carbon content of phytoplankton that became sequestered on the ocean floor proved to be disappointingly small. The trials are continuing, but many scientists fear that this approach to slowing global warming can make only a minor contribution to the growing crisis. And there could be unintended negative environmental consequences as well.

Dinoflagellates are also critically important symbionts of marine animals. One group of them forms mutualistic symbioses with such marine animals as jellyfish, sea slugs, coral, and certain giant clams. Via photosynthesis, the dinoflagellate supplies the animal with organic nutrients and the animal supplies the dinoflagellate with a place to live. The association between dinoflagellates and coral is particularly sensitive to environmental stress, as we will see in the next chapter. And this stress also contributes to the release of DMS.

The symbiosis between dinoflagellates and certain giant clams (species of *Tridacna* such as *T. gigas*) has drawn particular attention because of its economic importance. These clams, which are native to the shallow coral reefs of the South Pacific and Indian oceans, reach the enormous size of four feet across and can weigh up to 500 pounds. They reputedly live for over a hundred years. Of course, such huge creatures spawn legends: attacking, capturing, and drowning swimmers and producing gigantic pearls weighing as much as fourteen pounds. The stories about pearls might be true, but those of the aggressive killer clam are not plausible. The clam is fixed on the ocean floor, and although the shell opens during the day to expose its dinoflagellate-laden mantle to sunlight, it closes at night or when touched. It closes quite slowly, hardly fast enough to clamp shut on any but the most lethargic swimmer. But if it did close on a foot, perhaps, it could certainly hold the swimmer underwater.

The clam has been consumed for food by native peoples and is still a highly prized delicacy, called *himejako* in Japan. As a result, its populations have been decimated, and the species is now considered endangered. Giant clams are attractive animals for aquaculture, and several Southwest Pacific governments are making substantial investments to develop clam-growing industries to serve an already-established market. Because their gills are rich in symbiotic dinoflagellates, giant clams do not have to be fed. They obtain most of their energy for growth from sunlight. These clams have been described as the only photosynthetic and therefore self-feeding farm animal known.

Such aquacultural ventures have encountered a significant complication, however. DMSP, which is produced by the clam's symbiont dinoflagellates, enters both the wild and farmed clam's tissues. DMSP itself is neither dangerous nor offensive. But the DMS derived from it confers a bad odor and taste, creating a distribution problem that will have to be addressed.

One class of microbes, dinoflagellates, and the DMSP they produce can affect the ocean environment, the weather above it, and Earth's climate as well. The following sightings will introduce us to microbes that change Earth's geology as well as its atmosphere.

Red Cliffs in Western Australia

There are many places in the world to make this spectacular microbe sighting, including the massive iron-ore deposits of Minnesota that quenched the industrial Midwest's thirst for steel, but perhaps the most beautiful location is in Karijini National Park of Western Australia.

Such attractive red rocks can be found worldwide, albeit less spectacularly, because the event that caused them about 2.5 billion years ago was a microbe-driven change in Earth's atmosphere. Oxygen-producing cyanobacteria became more dominant on Earth after about 2 billion years without oxygen, and the planet rather slowly acquired its oxygen-containing atmosphere.

Everything changed. Energetically generous aerobic respiration became possible, giving evolution a major new tool, which fashioned animals and eventually humans. But again, microbes led the way. They were the first to evolve aerobic respiration. We know this because all aerobic eukaryotic organisms, protists, plants, and animals owe their capacity for aerobic respiration to intracellular structures called mitochondria. And many kinds of evidence establish unequivocally that mitochondria are the descendants of bacteria captured long ago. The most powerful such evidence is molecular, but it is worth mentioning that the DNA which mitochondria contain is arranged in a circular structure, as it is in most bacteria. This DNA is inherited only maternally by animals because eggs contain mitochondria and the much smaller sperm do not. Mitochondria are also surrounded by a double membrane, just as gram-negative bacteria are. So eukaryotes' capacity for aerobic respiration is dependent on their keeping enslaved bacteria, as is plants' and algae's capacity for photosynthesis. A few protists are unable to respire aerobically, lacking mitochondria to do that work. The backpacker's nemesis *Giardia* is one. It is interesting to speculate whether these mitochondria-free protists are descendants of primitive eukaryotes that had not yet captured a premitochondrial bacterium or of more advanced forms that had lost them because they lived in an anaerobic environment where mitochondria would be a useless burden. Opinions about this dilemma have shifted back and forth. At the moment majority opinion seems to favor the theory that these protists are indeed the descendants of primitive eukaryotes that did not enslave a bacterium to do their respiration.

At first, the appearance of abundant oxygen in Earth's atmo-

sphere was an environmental disaster. Some have called it the holocaust for all then extant living things. Up until that time, life had
evolved in an oxygen-free environment. Although metabolically
useful, oxygen is a very toxic substance. It and its inevitable metabolic products, such as hydrogen peroxide (H_2O_2) and superoxide
(O_2^{2-}), destroy the molecular components of organisms. To tolerate the inevitable assaults of oxygen, all aerobes have evolved elaborate ways to protect themselves. These include producing enzymes
that catalyze the destruction of the most toxic forms of oxygen and
mechanisms of repairing the damage once it has been done—of
course, never with complete success. One theory of aging says that
in large part this incurable disease is the cumulative impact of
oxygen-caused damage.

Some strictly anaerobic microbes, today's descendants of our
planet's pre-oxygen environment that now persist in Earth's many
anaerobic niches, are instantly killed by the briefest exposure to air.
Laboratories where they are studied are provided with anaerobic
glove boxes, some even with anaerobic rooms.

Oxygen-evolving cyanobacteria caused this holocaust. The red
cliffs of Western Australia and red iron formations elsewhere are
evidence of it, and their location in geological strata tells us when
this momentous event occurred. Rocks record atmospheric changes
clearly and indelibly in Earth's open geological diary. One wonders
how our own impact on atmospheric carbon dioxide and climate
change will be written and who will read it.

We should not overlook, however, another less obvious but
extremely positive impact of an atmosphere containing oxygen,
namely the stratospheric ozone layer that developed from it. Without this protective blanket that absorbs ultraviolet (UV) light, Earth
would be a hostile environment. Some speculate that life exposed
directly to such UV-intense, unfiltered sunlight would not be possible. UV light is extremely destructive to cellular components, especially DNA. All aerobic organisms, therefore, have DNA-repair
mechanisms to deal with the damage that even today's lower UV

levels cause. Some are quite elaborate. One, for example, identifies the damage and its location, cuts out the damaged strand of DNA, and replaces it with functional DNA, using the opposite strand of DNA as a source of information. There is reasonable evidence that the selective pressure of DNA damage has even shaped the composition of DNA. In it, the nucleic acid base uracil (U), found in RNA, is replaced by less UV-sensitive thymine (T).

So the red rocks chronicle the history of the abundance of oxygen gas on our planet. Here on Earth there are two major kinds of iron-containing rocks—banded iron formations and red beds. Before we can use them to unlock the story of oxygen and its introduction, however, we must first review a small bit of inorganic chemistry.

There is lots of iron on Earth. It is the sixth most abundant element in the universe and the fourth most abundant on our planet. Iron exists in two oxidation states: plus two iron (Fe^{2+}), called *ferrous* iron and plus three (Fe^{3+}), called *ferric* iron. Compounds made of ferrous iron tend to be a pale green color and to be highly soluble in water. Ferric iron compounds in general are also soluble in water with the dramatic exception of ferric oxide, which is extremely insoluble. Ferric oxide, also called iron oxide or rust, is red. And ferrous iron is readily oxidized to ferric oxide by oxygen gas.

Banded iron deposits, the Earth's oldest geological forms of ferric oxide, were formed about 2.5 billion years ago, soon after cyanobacteria became dominant and started making significant amounts of oxygen. These deposits are rocks with, as the name implies, red, iron-rich bands separated by light-colored, iron-poor regions. They constitute a major source of iron ore. They were formed as sedimentary rocks when ferric oxide precipitated from shallow seas that were filled with ferrous iron, then abundant in Earth's oxygen-free, highly reduced environment. The reasons for the intermittent nature of this process, which caused the rocks to be banded, are not understood to the satisfaction of all geologists. Possibly during that era Earth experienced extremely cold ("snow ball") periods that stopped oxygen production by cyanobacteria or produced an ice sheet that prevented oxygen from coming in contact with ferrous iron in the seas. Banded iron deposits do not form today.

When banded iron deposits were forming, Earth's atmosphere remained almost oxygen-free because as quickly as oxygen was formed by cyanobacteria, it was taken up by ferrous compounds in the seas, forming ferric oxide. Earth's supply of iron had a tremendous capacity to use up oxygen. Today, greater quantities of oxygen are sequestered as ferric oxide than exist free in the atmosphere.

Then, a little over 2 billion years ago, the seas' supply of ferrous iron was depleted and newly formed oxygen began to spill over into the atmosphere, eventually reaching its present 21 percent with its panoply of positive consequences for enriching evolutionary potential and its disastrous impact on then extant organisms. The time of the buildup of oxygen in the atmosphere is recorded by the age of the first appearance of red bed iron deposits.

For sure, red iron deposits are a highly significant microbe sighting, and Western Australia offers a particularly beautiful view of them. But it is certainly not the only such sighting. Recently I saw a notable example of banded iron in Glacier National Park on the trail to Avalanche Lake. Iron deposits mark an important event in Earth's evolution and remind us of microbes' impact on its geology. The next sighting will introduce us to another way that microbes have sculpted Earth's geology.

A Visit to Carlsbad Cavern

Improbable as it might seem, Carlsbad Cavern, one of the most well known and frequently visited limestone caves in the collection of over eighty that make up Carlsbad Caverns in New Mexico, is the handiwork of microbes, as are many other limestone caves.

Carlsbad Caverns has been a National Park since 1930 and a World Heritage Site since 1995. Its principal cavern is best known for its size, beauty, and vast population of bats. It is over 1,500 feet deep and contains huge chambers that are richly decorated with stalactites, stalagmites, columns (where stalactites and stalagmites join), and other formations with such descriptive names as draperies, cave pearls, lily pads, and popcorn.

Carlsbad Cavern is home and sanctuary to about a million Mexican Freetail bats as well as lesser numbers of six other species. The Freetails spend their days packed tightly together, as many as 300 bats per square foot, protected from predators immobilized by their inability to hunt in the cavern's complete darkness. At dusk, seeking to gorge themselves on insects, the bats emerge from the cavern en masse and quickly, sometimes within about twenty minutes, form a spectacular, swirling mass. They then disperse on independent insect-hunting forays. After packing their stomachs several times with insects during the night, they return at dawn individually or in small numbers. The coming back, in its own way, is almost as spectacular as the leaving. As a bat flies over the mouth of the cavern, it folds its wings and plummets into the darkness, making a distinctive buzzing sound as it does. The bat show, at least for now, is largely limited to the summer months. In late October or early November, they head for richer bug territory in southern Mexico.

The general story of Carlsbad Cavern's formation has been known for some time. An understanding of its microbiological dependency, as we will see, came later. The cavern was carved out of a vast deposit of calcite (also called calcspar, a particular crystalline form of calcium carbonate, $CaCO_3$) that precipitated out of an inland sea. Subsequently, the sea evaporated and geological forces lifted the calcite reef up though overlaying salt and gypsum (hydrated calcium sulfate, $CaSO_4$) to just below Earth's surface. Then, cracks formed and hydrogen sulfide–containing water seeped into them. The hydrogen sulfide (H_2S) oxidized to sulfuric acid (H_2SO_4), which then reacted with the reef's calcite ($CaCO_3$), breaking it down to gypsum ($CaSO_4$) and carbonic acid (H_2CO_3). Carbonic acid is extremely soluble in water and gypsum is moderately soluble. Calcite is quite insoluble. So the water that continued to flow in washed the gypsum away as it formed, carving out a hole that eventually became a cavern. It was a slow process in our terms, only about two inches every hundred years, but over the vastness of geological time that is pretty rapid. Decoration occurred slowly, beginning during a wet period about a half million years ago, after the cavern was largely formed. Acidified water dissolved minuscule amounts of calcite and as it evaporated within the cavern, redeposited it in the beautiful shapes we now admire.

The vital role microbes play in the formation of limestone caverns has been revealed only relatively recently through studies on caverns that are still actively enlarging. These include the Kane Cave in Big Horn County, Wyoming; the Lechugilla Cave in New Mexico; and the Frasassi Cave in Italy. Microbes are directly responsible for the erosion of limestone, which creates fissures, sinkholes, and underground streams, producing a landscape that geologists call *karst*. But how? The crucial step is how hydrogen sulfide is oxidized to sulfuric acid. Until recently this process was considered to be spontaneous, or abiotic. Now it is clear that this is not the case. Microbes bring about this oxidation.

The microbes that do this are bacteria that fall into the class called *autotrophs* (self feeding), organisms that make all their PMs and hence organic constituents from atmospheric carbon dioxide. I have talked quite a bit about those autotrophs called photoautotrophs, such as cyanobacteria and plants that use light energy to drive their metabolism. The hydrogen sulfide–oxidizing bacteria found in these caverns and elsewhere are called chemoautotrophs because they belong to the class of organisms that obtain their metabolic energy from a chemical reaction. These are the same autotrophs we saw in the acid drainage from mines and in the manure pile. The name *chemoautotroph* means, although it is not specifically stated, that the energy-yielding reaction is an inorganic chemical reaction. Organisms called *heterotrophs* (other feeders), such as ourselves, obtain metabolic energy from an organic chemical reaction. We do not have to make our PMs from carbon dioxide because we have an organic compound available that we use for this purpose.

In carrying out the inorganic chemical reaction so as to obtain metabolic energy, the chemoautotrophic bacteria make the sulfuric acid as an endproduct, which enlarges the cavern. In so doing the bacteria grow, replicate, and produce biomass. They are the primary producers in the cavern's dark environment, just as cyanobacteria, algae, and plants are the primary producers in the majority of Earth's sun-drenched environments. As such they support the growth of other microbes that congregate as microbial mats within the cavern. They create their own complex microbial ecosystem that depends for its energy only on a source of hydrogen sulfate. The process does not need sunlight. In many respects, this ecosystem is similar to the microbe-supported one that flourishes near hydrothermal rifts on the ocean's floor.

Put in other terms, these microbial primary producers obtain metabolic energy through aerobic respiration of an inorganic compound, hydrogen sulfide. That means they need a source of oxygen as well as hydrogen sulfide, but in some waters within caverns the supply of oxygen is not that great. Microbes can solve that problem as well. They resort to anaerobic respiration using nitrate ion as an oxidant instead of oxygen: hydrogen sulfide (H_2S) and nitrate (NO_3^-) oxidize to sulfuric acid (H_2SO_4) and nitrite (NO_2^-), releasing metabolic energy.

Microbes build caves without benefit of sunlight or organic nutrients, using metabolic tools that only they possess. As they excavate, they provide nutrients for a complex underground ecosystem, but their greatest and most obvious impact is directly on Earth itself. Unsurprisingly, production of sulfuric acid and hence cave building goes on beneath our feet without sunlight, so seemingly crucial to life. Once thought to be environments hostile to most forms of life, caverns support a functional ecosystem thanks to the rock-carving and -building activities of microbes. Unsurprisingly, the once-distinct fields of geology and microbiology are now overlapping. The term *geomicrobiology* is becoming firmly established.

The Smell of Soil

Soil, particularly a freshly turned fertile soil, has a distinctive odor that can only be described as itself. Everyone is familiar with the odor of soil. It, too, is caused by a microbe. The odor itself is that of a rather simple twelve carbon atom, double-ringed structure called *geosmin* (which translates as "earth smell") with an astoundingly intense odor. The average person can detect as little as four nanograms (four trillionths of a gram) of it in a quart of air. That is a concentration of about 0.00052 parts per million on a volume basis, or somewhat less than half the concentration (0.0011 parts per million) at which humans can detect methyl mercaptan, the highly odoriferous compound added to natural gas so that we can detect leaks or be warned of an unlit burner on the kitchen range.

Some cyanobacteria and a few other organisms make geosmin as well, but the odor of soil comes largely from the Actinobacteria, more commonly known as actinomycetes, growing in it. These bacteria are readily isolated by spreading a diluted suspension of almost any soil on the surface of certain simple media in a petri dish. Most of the colonies that develop on the surface of such media are actinomycetes. They present themselves in a gorgeous variety of pastel shades of oranges, yellows, blues, and violets. They all smell like soil because they all make geosmin. As the name *actinomycete* ("radiating fungus") implies, they have certain funguslike properties. Their cells are long tubes, forming an interwoven mass that penetrates into a solidified medium and produces a layer of spores on its upper surface. Just by touching an actinomycete colony with a needle and noting its coherence, you can be assured that it is, indeed, an actinomycete, if assurance beyond its color and odor are needed.

Geosmin gives soil a pleasant odor, but it is rather disagreeable in drinking water and certain foods, including fish, beets, and even wine. Certain species of filamentous cyanobacteria have been identified as the culprits of adding geosmin and another objectionable, odoriferous compound, MIB (2-methylisoborneol), to water in storage reservoirs. Such contamination is a pernicious problem because these two compounds are undiminished by the usual treatments of drinking water.

Cyanobacteria are also largely responsible for the muddy, geosmin odor and taste of farmed freshwater fish such as bass, trout, sturgeon, and tilapia, as has been shown experimentally. I have purchased tilapia, which I usually like and appreciate for being ecologically friendly (as vegetarians they need not be fed other fish) that tasted as though it had been fried in a mud batter.

Geosmin is also alleged to give the "corked" odor (sometimes

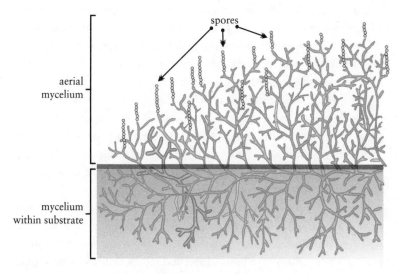

An actinomycete colony.

described as the smell of a damp basement) to wine—the reason
why the sommelier at an elegant restaurant offers the pulled cork
for your examination. The source is thought by some to be fungi
growing on the cork. Beets make their own geosmin, which tends
to be concentrated in their skin.

Still, actinomycetes are the notable producers of geosmin and are
largely responsible for soil's smelling the way it does. Why they
make it remains a mystery. But actinomycetes are experts at making unusual compounds. They are the major producers of antibiotics, a role which we have already discussed, and a good explanation
for why they make these lifesaving compounds is also lacking.

A Walk in Yosemite

A walk in Yosemite Valley or any other woodland scene is replete
with casual microbe sightings. They are good places to practice and
extend your microbe-watching skills.

One of the most dramatic sights for microbe watchers in Yosemite Valley is the many black streaks running down its majestic granite walls. These are cyanobacteria. Being photosynthetic, they are
able to derive their metabolic energy from sunlight and they obtain
carbon to make their PMs from carbon dioxide. And they are able
to fix atmospheric nitrogen as well. They need only a source of
minerals, which the granite supplies, and water, which runs down
these pathways in the spring. Then they prosper. Their resting cell
forms survive summer's heat and dryness along with winter's frigid
temperatures in order to resume growing the following spring when
water and warmth return.

Near a stream in the right season you will see lupine growing
in the sandy soil. If you examine the roots of one of them you
will discover nodules. If you cut open one of these nodules with a
pocket knife, you will note its blood-red interior, sure evidence that
nitrogen-fixing bacteria belonging to the rhizobial group are active

there, just as they were in the root nodules of legumes we discussed in Chapter 5. Close by, you are likely to notice alder trees, catkin-bearing relatives of birch trees. They, too, add to the environment's fertility by housing on their roots nitrogen-fixing bacteria belong to the genus *Frankia*. Even the black streaks of cyanobacteria on the canyon wall make a small contribution of fixed nitrogen.

Small lodge pole pines on the edge meadows march in to make a forest where there is now grassland. Some of these trees will bear the scars of last winter's snow and fungal attack. Blackened clusters of needles cover some of their lower branches, evidence of fungi having grown in a warm, humid chamber formed as snow melted first in this region. Last year's crop of deciduous leaves, now apparently completely gone, in fact have been reduced to their plant-nourishing nitrogenous and mineral constituents by a consortium of cellulose-degrading microbes—a mixture of fungi and bacteria working cooperatively.

In the late spring you will see mushrooms, the macroscopic fruiting bodies of ascomycete and basidiomycete fungi. Some might be boletes that an experienced mushroomer would rejoice to collect and eat, others poisonous and sometimes deadly amanitas. A pocket field guide and the company of an experienced collector will help you identify which are which. With a little research, any walk through a woodland or riparian environment can be rich with microbial sightings.

13

Closer to Us

Protists are usually defined as a sort of microbial leftover, constituting all eukaryotic microbes other than fungi and unicellular algae. The motile, nonphotosynthetic ones were once taxonomically separated and called protozoa. Although the name is still used, modern molecular methods of determining relationships among groups of organisms have discredited it. Some photosynthetic and nonphotosynthetic protists are quite closely related. In spite of their leftover status, protists are a fascinating group of microbes, both for their appearance and their activities. They are large enough, usually about ten times larger than an ordinary bacterium, that we can appreciate their cellular complexity using a common lab microscope. Some can be seen just barely with the unaided eye, challenging their good standing as microbes. Because all are clearly visible with an ordinary microscope—no staining required—we can view them alive and watch their fascinating activities as they swim, abruptly stop or turn, and sometimes consume one another. The creatures in the pond water you may have observed under a microscope in your high school biology class were largely protists. We share with them membership in the eukaryotic branch of the tree of life. Their evolutionary history is closer than prokaryotic microbes to our own.

We have already seen a number of protists in previous sightings, but in this chapter we will view some more that directly or indirectly impact our lives.

A Green and a White Coral Reef

If you have been snorkeling in almost any tropical waters—a place where it is relatively shallow (less than about 100 feet deep) between 30° north and 30° south latitude—you may have glided over a coral reef. The west coast of the Americas is an exception. The ocean there is quite chilly as a result of the upwelling of cold water from deep in the ocean and the cold-water ocean currents that sweep by it. Coral does not grow in water where the temperature falls below about 16°C (72°F). Coral is also largely absent from the west coast of Africa for similar reasons. If you have snorkeled around a coral reef, you may have noticed that they vary in color.

A white patch of coral is quite beautiful, more beautiful than a green one, but the coral in the white patch is ill, perhaps mortally so. It has lost its complement of nourishing symbiotic microbes— one or another species of dinoflagellates, the same group of eukaryotic microbes (protists) that lit up in the wake of a ship or that formed clouds.

Hard stony coral comes in many shapes, and their common names often reflect those shapes: cauliflower coral, lace coral, antler coral, brain coral, rice coral, and mushroom coral are a few examples. Soft coral can take on a variety of shapes. The color of coral also varies from yellowish through green and bluish to brown, depending on which particular dinoflagellate they harbor as well as the intensity and quality of light that they are exposed to. Dinoflagellates, as is the case with most phototrophic microbes, can vary their formation of accessory light-gathering pigments, and hence their color, to take full advantage of the light they happen to be exposed to.

The particular dinoflagellates that give healthy corals their color belong to a subgroup called zooxanthellae, which specialize in entering into mutualistic symbioses with marine animals. Certain species in the group are free living like most other dinoflagellates, however, and never live symbiotically. The various species that do also enter into symbioses with other marine animals such as clams, nudibranches, flatworms, sea anemones, mollusks, sponges, and jellyfish, as well as coral. In their free-living state zooxanthellae look like any other dinoflagellate with two flagella, one of which is wrapped around a groove in the cell's midsection, giving the microbe its characteristic pinched-in look. Living free, they obtain some of their nutrients by ingesting other microbes.

But when zooxanthellae enter into a symbiosis, they lose their flagella and their characteristic shape. They become undistinguished, oval-shaped cells totally dedicated to photosynthesis in order to nourish themselves and their host. The symbiosis of zooxanthellae with coral has become the focus of intense investigation in recent years because the vigor and even survival of coral depends on this source of nutrients. Coral has a profound influence on the ocean's ecology. It is viewed by oceanographers as an early-warning bellwether of damage to the ocean ecosystem. Like the canary in the mine, it is the first to show the effects of environmental degradation.

Individual corals, called *polyps,* are rather small, inconspicuous sea anemone–like animals. Each polyp is only a few millimeters in diameter and about two or three times as long. At their upper end, coral polyps display an array of stinging tentacles surrounding a central mouth. They use these tentacles to catch various kinds of phytoplankton and thereby to feed themselves. But the major part of a coral's nutrients are not found by the polyp itself. They are supplied by the dinoflagellate symbionts that live imbedded in their tissues. This symbiosis is mutualistic: it benefits the dinoflagellate as well as the coral. The coral provides a home and a high concentration of carbon dioxide to drive photosynthesis by the dinoflagellate, which in turn supplies the coral with nutrients, including carbohydrates and amino acids. Coral polyps live together in huge masses of genetically identical individuals that are interconnected by protoplasm-filled tubes (coenosarcs) through which they can exchange nutrients and their dinoflagellate symbionts as well. As the polyps of stony corals grow, they lay down an exoskeleton of rock-like, beautifully sculptured calcium carbonate ($CaCO_3$).

Over time these masses of coral polyps build giant reefs that have great ecological and even geographic impact. A coral reef provides a sanctuary for a vast and complex community of other creatures, including fish, algae, starfish, sea anemones, and, of course, microbes. Nitrogen-fixing cyanobacteria are particularly important members of the community. They contribute to the fertility and health of the reef by supplying utilizable nitrogen, an essential nutrient that is in short supply in otherwise nutrient-poor tropical waters. Other bacteria, as we will see, have a negative impact on the reef. They attack and sometimes kill the coral's dinoflagellate symbiont, the reef's vital nutritional engine.

Because corals are restricted to relatively shallow water in order satisfy their symbiotic dinoflagellate's need for light and their own need to remain submerged, coral reefs typically form off the shores of tropical islands, creating lagoons between reef and shore with their own special ecology. In certain places in the Pacific Ocean the island itself has subsided, leaving only a circular reef surrounding an empty lagoon. Such islandless reefs, called atolls, include Bikini, Midway, and Kwajalein, made into household names by events during World War II and its aftermath. Kwajalein, which has a sixty-mile-long lagoon and is the world's largest atoll, was the site of the battle in January 1944 in which all but fifty-one of its 3,500 Japanese defenders were killed.

Barrier reefs have the same relationship to landmasses as atolls do to islands. As the landmass subsides and continues to subside, large channels are formed between the reef and the mainland. The largest such reef is the Great Barrier Reef off northern Australia. It is 1,240 miles long. Although it may seem that such an enormous reef must be very old, the world's coral reefs are less than 10,000 years old—all have formed since the last ice age. Not all reefs are continuing to grow, however. Some reefs appear gray or white and no longer teem with life.

The threatened white coral has undergone a process called *bleaching* in which the coral polyp actively expels its dinoflagellate symbiont from the pockets it occupies on its host's surface. Any of a number of environmental stresses can diminish the dinoflagellate's photosynthetic productivity and, as a consequence, trigger the polyp to get rid of its food-supplying partner by expelling it from its home in the polyp. Reefs near urban population centers and farming regions are stressed by the pollutants and sediment flowing downstream into them. The increased intensity of UV light resulting from Earth's thinning ozone layer also contributes to coral's stress.

Overfishing can destroy a coral reef by depleting the reef of its cadre of vegetarian fish, which graze on the algae growing on the surface of the coral. With fewer fish, algae accumulate, shading the coral and depriving the dinoflagellate of its ability to photosynthesize effectively. If it cannot produce food, the coral polyp evicts it. Small tropical fish of the sort sold for household aquaria as well as certain fish caught for human consumption are important contributors to algal cleanup on the reef. And both groups have been massively overfished in recent years, sometimes in highly wasteful ways. For example, it has become relatively common to add cyanide to the waters around reefs to narcotize tropical fish so they can be gathered easily. Many fish do not survive such treatment.

But warming of ocean waters appears to be the major factor that damages the dinoflagellates, causing coral to expel them and bleach. Whether a consequence of temporary temperature spikes caused, for example, by an El Niño event, or as the outcome of long-term global warming, a rise in ocean temperature of only a few degrees, even over a relatively brief period of a few months, can cause bleaching. Some bleached coral regain a dinoflagellate symbiont and recover, but if bleaching is prolonged, many coral die. Beginning in 1980, the frequency and extent of bleaching increased worldwide. Bleaching, along with pollution and disease, killed about 16 percent of the world's coral reefs between 1995 and 2005.

Powerful recent evidence supports the theory that the bleaching of coral is adaptive—that it has been evolutionarily selected for the coral's survival. That is, if the coral's symbiotic partner is not doing a satisfactory job of supplying nutrients, the coral rejects it, possibly later to acquire a better provider. In that way, the coral is able to adapt to a changing environment. The process appears to be highly evolved, even sophisticated, and undoubtedly has helped coral to survive previous environmental challenges. The adaptation depends on microbes' remarkable ability to evolve to withstand environmental insults. We have already encountered an example of this skill in microbes' rapid evolution to survive treatment with antibiotics. But the present array of environmental assaults that coral and their symbiotic partners are encountering may be occurring too rapidly and extensively for this adaptive scheme to be successful. The accelerating rate of environmental change could exceed the evolutionary capacity of coral and dinoflagellates to adapt.

An international consortium of scientists recently concluded that coral reefs would change rather than disappear entirely. Still, damage to coral reefs has been and will continue to be massive. The same group of scientists concludes that reefs are in serious decline worldwide and that there are no pristine reefs left. Close to 60 percent of our remaining reefs may be lost by 2030. Saving coral reefs will depend on slowing or stopping global warming as well as instituting marine protective areas for fish, including no-take areas. These would afford a refuge from which coral larvae and adults could disperse to adjoining exploited areas.

Corals face yet another environmentally triggered threat. This one is microbiological. Weakened corals become susceptible to outbreaks of bacterial diseases such as blank band disease and yellow blotch or band disease. Yellow band disease, prevalent in the Caribbean Sea, is caused by species of the bacterial genus *Vibrio,* closely related to the bacterium, *Vibrio cholerae,* that causes the deadly human disease cholera. Curiously, this bacterium attacks coral's symbiotic dinoflagellate partner, bursting its cell walls (lysing it). It does not attack the coral itself.

Although white coral may appear to be beautiful, it portends a major threat to the ocean's bounty. We can only hope that microbial evolution, along with some helpful human intervention, will be sufficient to save the world's rich coral ecosystem.

Red Tide

Red tide is all too apparent during the summer months in many places on Earth, including both coasts of the United States. A large patch of water is covered by a heavy, colored growth, called a *bloom,* of phytoplankton. The colored water is a danger sign both for marine life and for human shellfish eaters. Some of the microbes found in a red tide produce powerful toxins that can kill fish in vast numbers. The danger of the sighting to humans is largely indirect. Filter-feeding mollusks including oysters, clams, scallops, and mussels, themselves unaffected, accumulate sufficient quantities of toxins to become damaging or lethal when eaten. This microbe sighting is the basis of the old adage of the northern hemisphere: never eat seafood during months that lack an "r" in their name. The warmer waters in the summer months increase the risk of seafood being poisoned. Perhaps the adage should extend to freshwater fish as well: toxic phytoplankton blooms are not limited to oceans. They also occur inland, creating similar havoc.

Red tide has been known and feared throughout human history.

It is clearly recognizable in the biblical description as one of the plagues of Egypt (Exodus 7, 20–21): "all the waters that were in the river were turned to blood. And the fish that was in the river died; and the river stank, and the Egyptians could not drink of the water of the river."

"Red tide," although a romantic descriptor, is in fact a misleading one. Both "red" and "tide" are inadequate. The majority of the phytoplankton that cause toxic blooms are dinoflagellates, and these microbes, as we noted when we discussed white and green coral, come in a variety of colors. Indeed, in addition to red, some blooms are green, orange, or brown. And not all affected waters are tidal. The term used by scientists, therefore, is *harmful algal bloom* (HAB), although this term certainly lacks the ominous ring. To call a red tide a harmful algal bloom has other difficulties: it identifies the microbes as algae that are now classified as protists. Moreover, other microbes, including diatoms and cyanobacteria, also cause toxic blooms. Only a few older scientists still refer to cyanobacteria as blue-green algae.

All these microbes in marine blooms are part of the ocean's phytoplankton. As we have seen elsewhere in this book, the microbial community of phytoplankton is vast, complex, and vital to Earth's ecology. Our own survival depends on them. Phytoplankton take up in their process of photosynthesis about half of all carbon dioxide utilized, so they are a vital component of the ocean's productivity and counter-balancers to climate change. Already, over 5,000 species of microbes have been identified in the phytoplankton and new ones are being recognized at a rapid rate. Only a few dozen of these have been implicated in HABs.

HABs, although varied in their effects, fall into one of two general types: those composed of microbes that do their damage by producing a toxin and those that do harm simply as a consequence of the massive numbers of them. The latter can sicken and kill some fish by plugging their gills. More devastatingly, they sometimes kill large numbers of fish by robbing an area of the ocean or an inland

water of its oxygen. Being composed of either photosynthetic protists or cyanobacteria, HABs produce oxygen when they are healthy
and sunlit. But at night, they stop producing oxygen and instead
use it up through respiration, sometimes depleting all of it. When
conditions worsen, the bloom dies, and respiring bacteria attack
members of the HABs; in so doing, they use up all ambient supplies
of oxygen.

Toxin-producing blooms need not reach such densities to be
harmful. Some are among the most highly toxic known substances.
They become concentrated by the filter-feeding bivalves that usually are the vehicle for transmitting them to humans, although
sometimes humans are poisoned by swimming, ingesting, or merely
being exposed to the spray of HABs. Marine mammals such as
whales, sea lions, and sea otters are also affected, as are certain
birds.

HAB-associated illnesses are classified by the principal symptoms
that the toxins cause: neurotoxic, paralytic, diarrheic, and even
amnesic. Particular microbes produce particular toxins and their
blooms tend to recur in the same locations. For example, one red
tide caused by a bloom of the dinoflagellate *Gymnodinium breve,*
which produces brevitoxin A, occurs on the Florida Gulf Coast
whenever conditions are favorable for it. More widespread in the
Caribbean, Hawaii, and Guam is ciguatera fish poisoning, caused
by various armored dinoflagellates releasing ciguatoxin and maito-
toxin. Northern regions might be affected by dinoflagellates that
release saxitoxin, causing paralytic shellfish poisoning; or certain
diatoms (*Pseudo-nitzschia* spp.) might release domoic acid, causing
amnesic shellfish poisoning in its victims. *Dinophysis* spp., dinoflagellates, may spread their toxin that causes diarrhea, okadaic acid,
worldwide.

A probable reason for recurrence in the same location is the formation by dinoflagellates of resting cells called *cysts.* Late in the
season they form sexual cells (gametes) that mate and produce
cysts. These sink to the bottom, forming a biological reservoir of

the microbes, which germinate and produce subsequent blooms when conditions are again favorable. There is little doubt that the frequency and intensity of HABs has increased in recent years, probably for a combination of reasons. Global warming certainly contributes, as does increased nitrogen- and phosphate-rich runoff from farmlands and regions of human habitation.

We have already had many encounters with two of the major groups of microbes (dinoflagellates and cyanobacteria) that cause HABs, but not the third (diatoms). They are one of the most abundant groups constituting phytoplankton. The distinguishing feature of diatoms is their unique and beautifully sculptured cell walls, which are composed of silica (hydrated silicon dioxide, SiO_2), the principal ingredient of glass. Silicon is the most abundant element on Earth, and its oxidized form, silica, is found everywhere in nature. It is the major constituent of sand in inland and nontropical oceans. Minerals such as quartz, flint, and opal contain silica. But silica is a relatively rare component of biological structures; dia-

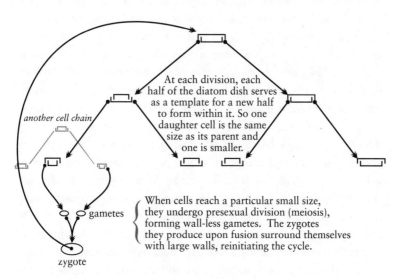

another cell chain

At each division, each half of the diatom dish serves as a template for a new half to form within it. So one daughter cell is the same size as its parent and one is smaller.

gametes

When cells reach a particular small size, they undergo presexual division (meiosis), forming wall-less gametes. The zygotes they produce upon fusion surround themselves with large walls, reinitiating the cycle.

zygote

The curious life cycle of a diatom.

toms, horsetail plants *(Equisetum)*, and certain sponges are notable exceptions.

The silica wall of diatoms consists of two separate parts, called valves or *frustules* (from the Latin "cut in half"). These frustules fit together like the two halves of a petri dish or a pillbox, one overlapping the other. Replication by division of a diatom cell encased within a nonexpandable silica wall presents a confounding challenge that is solved in an unexpected way. During cell division, each half of the wall serves as a template to make a new matching half that fits within it. Of course, this means that with each division the average size of progeny cells becomes smaller and smaller. When cell size reaches a certain functional limit, the process starts all over again. Some cells escape from their silica enclosure to become gametes, which fuse to produce zygotes. These zygotes form new, larger, petri dish–like cell walls around themselves.

Diatoms are heir to all the exigencies of living things. Other creatures consume them, and they succumb to myriad growth-arresting or lethal events. When they die, they decay as all the others do. But their cell walls are almost indestructible. These accumulate on the ocean floor and the bottoms of inland bodies of water in huge layers, which over geological time are distributed by the movement of tectonic plates over Earth's continental landmasses. Such deposits of diatom walls, variously known as *diatomaceous earth, diatomite,* or *kieselgur,* constitute an important article of commerce. Among its many uses is the making of dynamite, an application devised by Alfred Nobel in the 1880s that made him rich, enabling him to finance the world's most prestigious prizes for human achievement. Dynamite is made by absorbing highly explosive, unstable, and dangerous nitroglycerine into diatomaceous earth, making it stable and therefore transportable. Perhaps because it was so easy to use, it became a terrible new weapon in the labor unrest of the early twentieth century. Inexpensive, easy to store and transport, and relatively safe to use, it led to what some, comparing it to gunpowder's contribution to toppling feudalism, called "the cult of dynamite."

Another commercial use of diatomaceous earth was found by making a sweetened paste of it, which effectively controls cockroaches. In powdered form it controls other insect pests as well, including ants and bed bugs. Diatomaceous earth abrades and desiccates the integument of insects, killing them in an environmentally friendly manner. Insects do not develop resistance to diatomaceous earth and it is safe for animals, including humans. Largely nontoxic to mammals—it is even used as an additive to animal feeds to keep the feeds from caking—it can dry the skin and cause silicosis if repeatedly inhaled, however.

Human use of diatomaceous earth is not a recent innovation. Ancient Greeks used it as building material and as an abrasive. When diatomaceous earth is mined as blocks of stuck-together cell walls rather than being ground to a powder, it can be employed still as a building material or insulation. Large quantities of diatomaceous earth are also used in the manufacture of cement.

A noncommercial but esthetically delightful use of diatomaceous earth is an object of microscopic observation. It is only necessary to make a suspension of diatomaceous earth from filter aid, another product made from diatoms that is used to keep swimming pools clean, or from insecticide containing diatomaceous earth. Now look at it through the microscope. No staining or special preparations are needed. The rewards are rich. You will see the great variety of shapes diatoms can take—spindle-shaped, round, oval, crownlike, those that resemble sunbursts and those that look like intricate bracelets. The variations are almost endless.

Worldwide deposits of diatomaceous earth are massive, occurring in white, light beige, gray, or (rarely) black layers several hundred feet deep and several hundred square miles in area. Most mines are open pit, but some in China are underground. The United States is the largest producer, mining about 650,000 metric tons annually. China is in second place with about 350,000 metric tons. Most of the U.S. production comes from near the border between California and Nevada. The largest single producer is in Lompoc, California.

In spite of our massive use of diatomaceous earth, we are not about to run out of it because worldwide deposits are so huge. Diatoms have been growing on Earth for hundreds of millions of years and their cell walls are never degraded nor destroyed. As microbes go, diatoms have proven their usefulness to humankind, which cannot be said for the microbe we will discuss next.

A Child with Diarrhea, Stomach Cramps, and Sulfurous Bad Breath

This is almost certainly a sighting of *Giardia lamblia,* a protist (clinicians still call them protozoa) that has fascinated microbiologists since the beginnings of their science. It continues to fascinate because of its attacks on us and its own intriguing biology. Many of its cells are probably clinging to the wall of the child's upper small intestine. Why is it that *Giardia* is such a likely suspect? Diarrhea and stomach cramps have many causes, but the distinctive sulfurous bad breath, which is not always associated with the infection, tells the whole story.

Our knowledge of *Giardia*'s existence and association with diarrhea is almost as longstanding as the science of microbiology itself. While suffering a bout of diarrhea in 1681, Antony van Leeuwenhoek, a Dutch draper and amateur microscopist, perhaps even deserving the title of first microbiologist, discovered *Giardia* when he examined his watery excrement by squinting through his homemade, handheld microscope. Leeuwenhoek was marvelously curious. He had used his microscopes previously to look for sharp edges on pepper and for creatures in scrapings of his teeth and in canal water. Sharp edges, he concluded, were not the explanation for pepper's bite, but he found that his teeth scrapings and canal water were a fantastic zoo of "animalcules." This time, presumably, he was curious about the cause of his illness, and he was rewarded, as he reported in his refreshing, jargon-free style to the Royal Society of London, by seeing "tiny creatures, some of them a bit bigger, others a bit less, than a blood-globule but all of one and the same make. Their bodies were somewhat longer than broad, and their belly, which was flattish, furnished with sundry little paws, wherewith they made such a stir in the clear medium and among the globules, that you might even fancy you saw a woodlouse running up against a wall; albeit they made a quick motion with their paws, yet for that they made but slow progress."

There is little doubt that Leeuwenhoek saw *Giardia*. Its constantly beating flagella do cause rapid movements but only sluggish forward progress. Leeuwenhoek apparently associated these tiny creatures with his illness—all this some 300 years before Louis Pasteur announced the germ theory of disease. Leeuwenhoek's infection could not have been terribly unusual nor could it have materially affected his long-term health. He lived to be 91. Today the infection is still common. *Giardia* is the world's most often-detected intestinal protozoan parasite. Its prevalence is about 20 percent of the population of developing countries and up to 5 percent in others. The U.S. Centers for Disease Control and Prevention (CDC) estimates that there are between 100,000 and 2.5 million cases annually in the United States.

In spite of its recognized widespread prevalence, *Giardia* is undoubtedly blamed too often as a cause of diarrhea, particularly by backpackers in the mountains of the American West, who are encouraged to carry water filters to protect themselves from it. The poor personal hygiene many revert to while camping is most probably the greater culprit. Just suffering diarrhea in the mountains is not proof of *Giardia*. More likely it is caused by a bacterium, such as *Escherichia coli*, or a species of *Salmonella* or *Campylobacter* from a fellow camper. Robert Derlet, who extensively sampled lakes and streams in California's Sierra Nevada as a part of his twenty-year study of water quality, concludes that *Giardia* is so rare in this environment that backpackers, on average, would have to drink 250 gallons a day to become ill. State departments of health agree. A survey of forty-eight of them found that only two considered *Giardia* a problem for backpackers, and those two offered no data to support their concern. So, buying a water filter and carrying it to the mountains benefits the manufacturer, but it may be unnecessary for avoiding *Giardia*. Water near large numbers of campers or animals is a different story, however.

Giardia is distinctive in many respects, including its appearance,

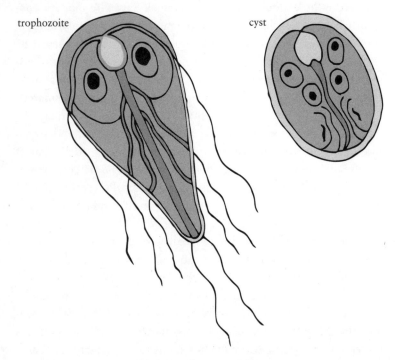

Giardia trophozoite and cyst.

which alone is sufficient for positive identification. *Giardia* resembles a pear cut in half through its long axis, with two symmetrically located nuclei giving it an eerie, facelike appearance. It is mobile by reason of its four pairs of flagella (the paws that Leeuwenhoek spoke of), and on its underside there is a concave disc, sometimes called a sucking cup, with which it can attach itself firmly to the intestinal wall so it is not swept away with the intestine's constantly moving contents.

In its well-nourished, secure environment, *Giardia* multiplies by cell division. In response to some still-mysterious environmental signal, however, division stops, and this distinctive, strangely beautiful cell, a trophozoite (Greek for "motile feeder") develops into a plain-looking, thick-walled cyst, which is passed in feces. Cysts are clearly resting forms. But the few studies that have been published suggest that they do not survive for very long in a natural environment, particularly one exposed to sunlight. They settle quickly in water, so that the top eighteen inches of an undisturbed mountain lake is likely to be free of *Giardia*. When cysts are ingested, stomach acid stimulates them to germinate and again become trophozoites. They are the form that multiplies and they do so by dividing, but only if the cells are attached to a surface. *Giardia* is not known to have a sexual phase of its life cycle.

Curiously, the reason that *Giardia* causes diarrhea as well as why it sometimes causes a telltale sulfurous breath remains a mystery. It does not produce a toxin. Just clinging to the intestinal wall, perhaps covering an area sufficient to interfere significantly with the intestine's absorptive responsibilities, is sufficient to cause diarrhea.

Giardia trophozoites look complex, but their structural composition is quite simple for a eukaryotic cell. They do not contain mitochondria or the other intracellular structures (for example, endoplasmic reticulum, golgi, and lysosomes) that almost all eukaryotic cells contain. Their lack of mitochondria has attracted the special attention of biologists because it carries evolutionary implications.

As I noted previously, overwhelming evidence shows that mitochondria are the present-day remnants of bacteria captured early in the evolution of eukaryotes, which makes eukaryotes, including us, capable of aerobic respiration as a means of generating ATP. Other evidence shows that *Giardia* is a terribly primitive eukaryote, located on a deep, early branch of the tree of life.

The tree of life, analogous to a family tree because it shows relationships by descent, is somewhat bushlike. Those branches such as the one bearing *Giardia* that form near the base are presumed to be primitive. They proceeded on their own line of descent before the common stem had become more highly evolved (see, for example, the phylogenetic tree illustrated in Chapter 1).

It seems likely, therefore, that *Giardia* and other primitive eukaryotes in this branch might be descendants of early eukaryotes in a line that had not yet captured these bacteria to perform aerobic respiration for them. But that theory is now open to question because of the discovery that the *Giardia* genome does contain genes that came from mitochondria. So possibly *Giardia's* ancestors once had mitochondria but lost them. These genes, however, might have come from other microbes instead of directly from mitochondria.

Transfer of genes from one organism to a completely different one has over the past decade been shown to be more common than was once expected. It confounds studies on evolution because it opens the possibility that an organism's genes or characteristics might not all be derived linearly from its hereditary past. They might have been even relatively recently acquired from a different hereditary lineage. Quite clearly such *lateral* or *horizontal gene transfer*, as it is called, adds a major complication to deciphering evolutionary lineages.

Giardia is not the only mitochondrion-free protist that causes outbreaks of diarrhea. Another one, *Cryptosporidium*, is also a major troublemaker. It causes a diarrhea that usually lasts one or two days, but it can continue for as many weeks. For those with weakened immune systems, such as HIV/AIDS, cancer, and transplant patients as well as those with inherited diseases of the immune system, a *Cryptosporidium* infection can be life threatening. *Cryptosporidium* is particularly pernicious for two reasons. First, it is highly infectious. Fewer than ten cells, possibly only one, can cause an infection. It is also highly resistant to chlorination, the principal treatment that we depend on to protect our water supplies. Huge numbers of infectious cells are shed in the feces of an infected individual.

Cryptosporidium infection is relatively common in the United States. About 2 percent of the population carries the protist, and antibody studies indicate that 80 percent of us have had the disease. The first recognized case of *Cryptosporidium*-caused disease occurred only in 1976, when a three-year-old girl from rural Tennessee suffered gastrointestinal disease for two weeks. A major outbreak of *Cryptosporidium* infection occurred in Milwaukee, Wisconsin, in April 1993. More than 400,000 people were infected and over 4,400 of them hospitalized. The illness was severe. It lasted on average twelve days, and at the peak of their illness patients had an average of nineteen watery stools a day. Several immunocompromised persons died. The source of the outbreak was traced to the South Milwaukee Waterworks plant; 50 percent of the people it served became ill. Chlorination procedures at the plant were suspected, but they were properly adding recommended levels of the disinfectant. The thick-walled oocysts that *Cryptosporidium* produces, however, are able to withstand such treatment. Why, then, had the water supply previously been safe, and for that matter, why are any water supplies safe from *Cryptosporidium*? All waterworks plants depend on chlorine for antimicrobial protection.

The answer is that water suppliers had unknowingly depended on the filtration procedures used to clarify water supplies to remove *Cryptosporidium* oocysts as well. Although usually filtration would be sufficient, in this case the clarification failed to function optimally and the outbreak was an unexpected consequence.

Although most microbes mean us no particular harm, a few can make us quite sick. *Giardia* was one of the first microbes so recognized and described, and the related microbe *Cryptosporidium* is one of the most recently recognized to cause disease. Primitive protists that lack mitochondria have fascinated since microbiology's beginnings and they continue to do so. They are spectacular to look at, they pose fundamental evolutionary questions, and they present manifold real and imaginary health concerns.

14

Survivors

Microbes inhabited this planet long before we did, and they may continue to do so if some cataclysmic event were to eliminate us. Of course, their survival would depend on the nature of the cataclysmic event. If it were intense radiation, many most probably would survive.

A String of Mushroom Clouds

Here is one sighting we all hope we will never see, other than on an old recording of the movie *Dr. Strangelove*. A catastrophic, planetwide rain of intense radiation resulting from a series of atomic explosions would eliminate many extant species, including our own. But some species would probably survive in a radiation-drenched world. Certainly, microbes would be among them. Certain microbes, most notably the bacterium *Deinococcus radiodurans,* are extraordinarily resistant to radiation. Whereas a radiation dose of 500 to 1,000 rads is sufficient to kill the average human, *D. radiodurans* is not harmed by ten thousand times that much— 5 million rads.

Without question, widespread intense radiation would cause mass extinction. But mass extinction is not without precedent. The fossil record reveals that at least five such extinctions have occurred during life's tumultuous history on Earth, and the recording of each of them is abundantly clear in the fossil record. Above a certain geological layer, a moment in geological time, we find dramatically less diversity of fossils than below it. Such a difference in strata represents the before and after of mass extinction. The missing species in the higher, more recent layer became extinct as a consequence of some catastrophe. The P-Tr extinction, the most mammoth of these cataclysmic events that we know of, occurred some 250 million years ago. It marks the transition from the Permian to the Triassic in our division of geological ages. It eliminated 96 percent of Earth's marine species and over 70 percent of those living on land.

The most well known and plausibly explained extinction event, though of lesser impact, is the K-T extinction, which occurred 65 million years ago as the Cretaceous age ended and the Tertiary geological age began. It eliminated most dinosaurs, except for those that gave rise to birds, along with almost all of the large invertebrates on land and sea. The explanation offered by Luis Alvarez, a Nobel Prize–winning physicist, and supported by evidence accumulated by his son Walter and his colleagues is now generally accepted by geologists and biologists. Alvarez suspected that a meteor or small asteroid was the culprit because the transition boundary layer was heavily laced with iridium, a rare element on Earth but a much more abundant one in meteorites. Others suggested that the extraterrestrial missile must have been huge, some six miles in diameter, and that it struck somewhere in North America. Later, satellite imagery revealed the actual impact site: an egg-shaped crater some 100 miles across torn into the Yucatan Peninsula, then covered by a shallow sea. The impact caused a period of intense heat locally, followed by an extended worldwide period of chilling and decreased sunlight due to atmospheric debris. Although the K-T geological layer is thin, the dying off of species probably continued over an extended period.

In contrast, lethality following intense radiation would be relatively rapid. Almost all organisms, including microbes, would be susceptible, *Deinococcus radiodurans* being a notable exception. This curious bacterium is extremely resistant to gamma radiation, the principal destructive radiation of a nuclear explosion, as well as to ultraviolet radiation and prolonged desiccation. The details of the extraordinary resistance of *D. radiodurans* are not completely clear, but the general outlines are. The primary cellular target of lethal gamma radiation is DNA: it kills a cell by breaking its DNA into fragments. Most cells are particularly sensitive to chromosomal breaks. *Escherichia coli,* for example, cannot survive even two or three complete (double-strand) breaks. In contrast, *D. radiodurans* is unaffected by radiation sufficient to cause hundreds of them.

D. radiodurans does produce enough carotenoid pigments (compounds known to offer screening protection against radiation) to color its cells an intense red, and its cells do have particularly thick walls, but *D. radiodurans* does not resist high levels of radiation by preventing radiation from entering its cells and causing chromosomal breaks. Instead, it is highly skilled at repairing the breaks once they form. To do this, *D. radiodurans* carries several copies of its chromosome in each cell, and it uses a piece of an unbroken segment that overlaps a break as a patch to repair the damaged area. Special enzymes adept at cutting and resealing DNA make the patch, which repairs the break precisely. The patched DNA is indistinguishable from that of an unbroken chromosome. In addition, a cell of *D. radiodurans* can call on adjoining cells for an extra reserve of patching material, if the supply within a single cell runs out. Four cells of *D. radiodurans* stick together, forming tetrads, and following radiation, DNA from one cell of the tetrad migrates through passages within cell walls into an adjoining cell.

D. *radiodurans'* extraordinary resistance to radiation presents an evolutionary conundrum. It can withstand higher levels of radiation than are present on Earth to act as a selective pressure. Evolution isn't proactive—it cannot act presciently to account for humans' not-yet-acquired skills and foibles. The best scientific guess for why D. *radiodurans* acquired its high-level resistance to radiation is that such resistance is a byproduct of selection for surviving prolonged, severe desiccation, a condition that also breaks DNA into fragments. Some support for the hypothesis that desiccation was the actual selective pressure comes from the fact that D. *radiodurans* is found in the extremely dry valleys of Antarctica as well as radiation-intensive venues such as irradiated canned meat and medical equipment.

An alternative explanation is that D. *radiodurans* or its antecedents evolved on primitive Earth when it was subject to much more intense radiation. There is some support for this hypothesis as well. The tree of life devised by modern molecular methods does, indeed, indicate that D. *radiodurans'* origins are ancient. The branch leading to it separates near the tree's root.

Of course, other human-made or natural catastrophes might lie in Earth's future. In another billion years or so the planet will continue to heat as the sun becomes brighter; collision with an asteroid of the sort that led to the demise of dinosaurs could cause a long and lethal winter. Microbes are better able to cope with all such environmental hostilities. I began this book by noting that microbes appeared on Earth 3.5 billion years before we did. It is only reasonable to expect that they will be here, whatever Earth's condition, long after we are gone. It seems that Louis Pasteur may have been right: microbes, indeed, will have the last word.

Glossary

Acknowledgments

Index

Glossary

μ: Greek letter "mu," used to designate micro, meaning millionth. For example, μg, designating microgram, is one millionth of a gram; μm, one millionth of a meter; and μL, one millionth of a liter.

acellular: Not made up of cells.

acetate: Neutralized acetic acid, vinegar (CH_3-COO^-).

acetogenic: The capacity to make acetate.

acid mine drainage: Acidic effluent from mines produced by the oxidation of pyrites or other reduced sulfur by microbes.

acidophile: An organism that thrives at a low pH.

acid rain: Rain with a low pH owing to its content of sulfuric acid.

actinomycete: Representative of a group of bacteria that grow as masses of fungallike tubular cells.

aerobic respiration: The cellular process by which nutrients are oxidized at the expense of oxygen while generating ATP.

amino acid: Building-block monomers of proteins; twenty occur naturally in proteins.

ammonification: The microbial process by which ammonia is derived from nitrogen-containing organic compounds.

anaerobic: In the absence of oxygen.

anaerobic respiration: The cellular process by which nutrients are oxidized at the expense of a compound other than oxygen, and ATP is generated.

anammox: The microbe-mediated anaerobic oxidation of ammonia as a result of its reaction with nitrite.

anoxic: Synonym of anaerobic.

anthrax: Disease caused by *Bacillus anthracis*.

antibody: Proteins made by the immune system that bind to antigens.

antigen: A compound, usually a protein, that elicits the formation of antibodies.

Archaea: One of two biological domains consisting of prokaryotic microbes. The other is Bacteria.

ascomycete: Any of 2 groups of fungi that forms ascospores in an ascus.

assimilation: The incorporation of a nutrient into the components of an organism.

atmosphere of pressure: The pressure exerted by Earth's atmosphere (14.7 pounds per square inch).

ATP: Adenosine triphosphate. The compound in which energy derived from energy-yielding reactions is stored for use in driving reactions that require energy.

ATP synthase: The cytoplasmic membrane-imbedded enzyme through which protons driven by a concentration gradient flow into the cell generating ATP.

attenuation: The lessening of a pathogen's power to cause disease.

autolyze: Self digestion and disintegration of a microbe.

autotroph: An organism that derives its organic constituents from carbon dioxide.

Bacteria, sing. bacterium: One of two biological kingdoms comprising prokaryotic microbes. The other is Archaea.

bacteriophage: A virus that attacks bacteria, commonly called a phage.

basidiomycete: Any of 2 groups of fungi characterized as forming basidiospores on a basidium.

beta carotene: The most common form of carotenes, naturally occurring pigmented compounds that are particularly abundant in plants.

bioelement: Any of the six elements—carbon, hydrogen, nitrogen, oxygen, phosphorus, and sulfur—comprising the macromolecules of all cells.

biofilm: A collection of microbial cells growing together as a film.

bloom: A dense growth of photosynthetic microbes.

blue-green algae: Former name of cyanobacteria.

capsule: The thick gelatinous layer, usually composed of polysaccharide, that surrounds some microbial cells.

carotenes: A class of naturally occurring pigmented compounds that are particularly abundant in plants.

carotenoids: Carotene-like compounds.

catalysis: The speeding mediated by catalysts of a chemical reaction by lowering its activation energy barrier. *See* enzyme.

cecal digesters: Animals such as horses and rabbits with an enlarged cecum that functions much like a rumen.

cell membrane: The thin structure that surrounds all cells.

cellulase: The enzyme that hydrolyzes cellulose.

cellulose: A polysaccharide made of glucose monomers joined by beta linkages.

chemiosmosis: The generation of ATP by forming an ion gradient across a membrane which then flows through membrane-imbedded ATP synthase.

chemoautotroph: An autotroph that obtains metabolic energy (ATP) from an inorganic reaction.

chemotaxis: The movement of motile microbes toward a chemically favorable environment.

chloroplast: The intracellular organelle of plants and algae derived from an anciently captured cyanobacterium where oxygenic photosynthesis takes place.

clone: (noun) The genetically identical progeny of a single organism or cell. (verb) The process of obtaining a set of genetically identical organisms or genes.

conidia, sing. conidium: Asexual spores formed by certain microbes.

constitutive: Descriptor of enzymes that are always formed (need not be induced).

coprophagia: Consuming feces.

cyanobacteria, sing. cyanobacterium: A class of bacterial phototrophs that produce oxygen as plants do as a byproduct of their photosynthesis.

cytoplasm: All the contents of a cell inside its cell membrane, other than nuclear material.

denature: Destroying a protein's three-dimensional structure and, thereby, its activity.

desulfurylation: The removal as H_2S of sulfur atoms from organic compounds.

diatom: Any of 2 groups of phototrophic protists with cells surrounded by silica walls.

differentiation: The process by which cells change their appearance during development.

dinoflagellate: A photosynthetic protist characterized by having two flagella, one of which is wrapped around the cell body.

diploid: A cell containing two sets of chromosomes (*see* haploid).

disaccharide: A sugar composed of two chemically linked simple sugars (monosaccharides). Sucrose and lactose are examples.

double helix: A synonym for DNA based on its twin helically intertwined single-stranded structure.

endospores: Highly resistant resting cells formed inside the cells of certain bacteria. *Compare* exospores.

endosymbiont: A microbe living inside the cell of another organism.

enology (oenology): The science of winemaking.

enzyme: A protein that catalyzes a biological reaction. A few biological reactions are catalyzed by RNA molecules called ribozymes. Catalysts speed the rates of reactions. Enzymes end with the suffix *-ase*.

ergot: The common name for the toxin (ergotamine) produced by the fungus *Claviceps purpurea*.

ethanol: A synonym for ethyl alcohol (CH_3-CH_2-OH).

Eucarya: The biological domain that includes all eukaryotic organisms, i.e., plants, animals, fungi, and protists.

eukaryote: An organism with eukaryotic cells, i.e., belonging to the Eucarya domain.

eukaryotic: The complex cell anatomy characteristic of those of all organisms other than bacteria and archaea.

eutrophication: The dense growth of phototrophic microbes that occurs in bodies of water overly enriched with nitrogen or phosphorus.

exospores: Resting cells formed on the exterior of the producing cell.

extremophiles: A microbe that flourishes in an environment that is hostile to most organisms.

fermentation: The anaerobic process, other than anaerobic respiration, by which some microbes generate ATP.

fermenter: A large tank, usually stirred and aerated, in which mass quantities of microbes are cultured.

fistula: An artificial opening. The commonly used term to describe an opening from the rumen to the environment.

flagellum: A cellular appendage of locomotion. Those on prokaryotes rotate; those on eukaryotes move with a whiplike motion.

fungi, sing. fungus: A group of eukaryotic microbes, characterized in the main by long tubular cells.

gametophyte: The growth phase of a plant that produces gametes.

genome: The total genetic complement of an organism.

geochemical: Combination of geological and chemical events.

gliding motility: The nonflagellar-mediated movement of some microbes on a solid surface.

glycogen: A branch-chain polysaccharide made of glucose monomers joined by alpha linkages. Starch is the straight-chain form of this polymer.

gram-positive: One of two principal cell types of bacteria that retain the Gram stain and lack an outer membrane.

gyre: A circular or spiral form, a vortex.

halophile: A microbe that flourishes in an environment containing high concentrations of salt.

haploid: A cell containing a single set of chromosomes (*see* diploid).

hemicellulose: A mixed polysaccharide that is a major constituent of plant cell walls.

hemoglobin: An iron-containing red pigment that binds and releases oxygen. Some forms bind and release CO_2 and H_2S as well.

hemorrhagic: Characterized by bleeding.

heterocyst: The thick-walled, nitrogen-fixing cells produced by certain cyanobacteria.

heterotroph: An organism that uses organic compounds as its major nutrient.

hydrostatic: The forces and pressures exerted by water.

hyphae, sing. hypha: The tubelike cells of fungi or other microbes with cellular structures similar to those of fungi.

inducible: A descriptor of enzymes that a microbe makes only when needed; for example one made to metabolize a compound only when it is available.

inoculum: A small quantity of microbial cells added to initiate growth.

isotope: A form of an element with the same chemical properties but a particular atomic weight, sometimes designated by a preceding superscript, for example ^{14}C, a radioactive isotope of carbon.

legume: A large family of plants with bonnet-shaped flowers, which usually form nodules on their roots that are populated by nitrogen-fixing bacteria.

lichen: A symbiosis between a fungus and a phototroph (either an alga or a cyanobacterium).

lignin: A component of woody plants that is only slowly degraded by microbes.

lithotrophs: A synonym for chemoautotrophs; microbes that utilize CO_2 as a source of carbon and oxidize an inorganic compound to generate ATP.

macromolecule: Literally, any large molecule. Used in biology to refer collectively to those large molecules found in cells, namely: proteins, nucleic acids (RNA and DNA), polysaccharides, and some lipids.

malolactic fermentation: The fermentation that occurs in some wines by which malic acid is converted to lactic acid and carbon dioxide.

Mediterranean climate: A climate characterized by rainy winters and dry summers.

meromictic lake: A lake that does not "turn over"; upper and lower regions do not mix.

metabolism: All the chemical reactions that take place in a cell.

metagenomics: A culture-independent method of studying microbes in which DNA is collected, sequenced, and compared with that from known microbes.

methane: Natural gas (CH_4).

methanogen: A microbe (all are archaea) that makes natural gas (methane). The adjectival form is *methanogenic*.

mitochondria, sing. mitochondrion: Microbe-derived intracellular organelles of eukaryotes that mediate aerobic respiration.

monomer: The unit building block of a macromolecule. For example, amino acids are the monomers of proteins and sugars are the monomers of polysaccharides.

mutant: A strain or individual that carries a mutation.

mutualism: A symbiosis in which both partners benefit.

mycelium: A network of tubular cells characteristic of most fungi and some bacteria (actinomycetes).

neurotransmitter: A compound that carries a nerve's signal from one nerve cell to the next.

nitrify: The conversion of ammonia to nitrate by microbes.

nitrogenase: The enzyme that fixes nitrogen.

nitrogen-fixing: The process of converting gaseous nitrogen (N_2) to ammonia (NH_3).

nucleic acid: A class of macromolecules made up of RNA and DNA.

oligosaccharide: A sugar made up of several simple sugars (monosaccharides).

organic acid: Any compound that can release a hydrogen ion (proton, H^+) is an acid. Organic acids usually contain carboxyl groups ($-COOH$), which readily release a hydrogen ion.

oxidation: The removal of electrons from a compound. Because electrons cannot exist alone, oxidations and reductions are always linked.

oxygenic: Producing oxygen gas.

pandemic: A widespread, sometimes worldwide, epidemic.

pathogenicity: Disease-causing capacity or ability.

pathway: A sequence of reactions through which a metabolic task, i.e., synthesis of a cellular component or utilization of a source of nutrients, is accomplished.

peptidoglycan: The generic term for a macromolecule composed of protein and sugars often used in place of the specific name, *murein,* the substance from which bacterial cell walls are made.

pH: A numerical index of acidity or alkalinity calculated as the negative logarithm of the concentration of protons (hydrogen ions, H^+). pH 7.0 is neutral and pH values decline with greater acidity. The approximate value of lemon juice is pH 2.0; pH 1.0 is the value of battery acid.

phage: *See* bacteriophage.

PHAs: Poly-beta-hydroxyalkonates, storage reserves that some bacteria make when nutrients are plentiful.

phosphorylation: The addition of a phosphate group (PO_4^{3-}) to a compound.

photoautotroph: An autotroph that derives its metabolic energy (ATP) from light and carbon atoms from CO_2.

photosynthetic: The capacity of an organism to obtain energy from light.

phototroph: An organism that derives its metabolic energy (ATP) from light.

phytoplankton: Literally "floating plant," a term used to describe a collection of waterborne microbial phototrophs that includes algae and cyanobacteria.

piezophile: An organism that flourishes at high hydrostatic pressures. Synonym for the older term, *barophile.*

pili, sing. pilus: Straight, stalklike appendages that extend from certain microbes. Synonym for *fimbri.*

plasmid: A small bit of dispensable DNA that many microbes carry.

PMs: Precursor metabolites. The twelve small molecules from which all cell components can be made.

polysaccharide: A macromolecule composed of chains of sugars. Glycogen, cellulose, and starch are examples.

probiotic: Administration of innocuous microbes to occupy the niches of destructive or pathogenic microbes and thereby exclude them.

prokaryotic: The simple cell anatomy characteristic of those of bacteria and archaea, which lack internal membranes and an organized nucleus.

protease: An enzyme that splits the bonds that link amino acids to form proteins.

protist: *See* protista.

protista: The formal term for protists: those eukaryotic microbes other than fungi.

proton: A hydrogen ion (H^+).

proton gradient: A concentration difference of protons across a membrane.

protozoa: A descriptive term for those protists other than algae.

psychrophile: An organism that flourishes at low temperatures.

pure culture: An organism growing by itself—axenic.

pyrite: Iron sulfide (FeS), also called fool's gold.

quorum sensing: The ability of certain microbes to sense and respond to the size of their population.

radiocarbon dating: Determining the age of biological material by measuring how much radioactive carbon (^{14}C) has decayed since it was formed.

reduction: The addition of electrons to a compound. Because electrons cannot exist alone, oxidations and reductions are always linked.

reverse transcriptase: The enzyme found in HIV and elsewhere that catalyzes synthesis of DNA from an RNA template.

rhizosphere: The region of soil surrounding a plant's roots.

RNA: The nucleic acid related to DNA that mediates protein synthesis and stores genetic information in certain viruses.

salt: The product of neutralization of an acid by a base. Sodium chloride (NaCl), the product of neutralization of hydrochloric acid (HCl) by sodium hydroxide (NaOH), is an example.

saltern: A facility for rendering sea salt.

smelting: Heating of mineral ores to purify them, often driving off sulfur as toxic sulfur dioxide gas (SO_2).

starch: A straight-chain polysaccharide made of glucose monomers joined by alpha linkages. Glycogen is the branch-chain form of this polymer.

sterilize: The elimination of all microbes from an environment.

sulfate-reducing bacteria: Bacteria that acquire metabolic energy by anaerobic respiration using sulfate as an oxygen substitute.

supercooling: Cooling water below its freezing point (0°C) without its freezing.

Superfund: U.S. government program charged with protecting people from abandoned contaminated waste sites.

symbiosis: Two organisms living together.

thallus: The body of a plant without leaves or stems, also of certain microbes.

toxin: A toxic substance made by an organism.

transduction: A mechanism of genetic exchange among prokaryotes, mediated by viruses.

transforming principle: Term used by F. Griffith to describe the DNA from pneumococci.

vaccination: The administration of an attenuated or killed pathogen or antigens from it to induce protective antibodies against the disease it causes.

virion: An intact virus particle.

virulence: The capacity to infect and cause disease.

yellow boy: Colloquial description of the yellow deposit that accumulates on the bottoms of streams fed by acid mine drainage.

Acknowledgments

Because this book rests on a career's worth of favorite microbe stories, I gratefully acknowledge my career's worth of teachers, students, and colleagues as well as my patient family. I'm especially grateful to those who played hands-on roles during this book's gestation, encouraging, advising, reading, finding errors and oversights: Tom Allen, Paul Baumann, Gary Hart, Jack Meeks, Fred Neidhardt, Nancy O'Connor, Elio Schaechter, and Mike Wolin. Special thanks are due Michael Fisher for his guiding hand and friendship, and very special thanks for Kate Brick's marvelously helpful contributions to the manuscript as well as her suspicious and therefore comforting eye, and most special tribute to the microbes themselves. They make this book and our existence possible; the world and our lives more interesting.

Index